AF540540

Seed Testing

Seed Testing

Dr. Mahesh Kumar

RANDOM PUBLICATIONS
NEW DELHI (INDIA)

Seed Testing

ISBN 978-93-5111-363-8

Published in 2014 in India by

RANDOM PUBLICATIONS

4376-A/4B, Gali Murari Lal, Ansari Road
New Delhi-110 002
Phone : +91-11-43580356, +91-11-23289044
e-mail: randomexports@gmail.com, sales@randompublications.com, info@randompublications.com

Reprinted 2021

Type Setting by : Keystoneprintads, Delhi-110051
Digitally Printed at : Replika Press Pvt. Ltd.

Preface

Seed testing is performed in dedicated laboratories by trained and usually certified analysts. The tests are designed to evaluate the quality of the seed lot being sold. The object of seed pathology testing is to determine the health of the seed lot for use in planting. Plant diseases are known to greatly reduce yield. Not all plant diseases are contained in or on seed but it is important to know if you have one or more that is. Seed may contain plant pathogens or agents that cause disease in plants. These diseases may affect storage, vigour, germination, market availability, harvest yield, seed appearance, or contain toxins.

Tests of the quality and other characteristics of seed need to be made at several stages during its progress from the parent tree to the seed bed. Tests for maturity and soundness of seed before and during collection in the forest and are usually associated with surveys of the abundance of the seed crop; they are intended to ensure that both the quantity and the quality of seed justify the effort and cost of collection. Several tests may be required at the seed processing depot, after extraction and cleaning and before the seeds are despatched to the nurseries or placed in storage. Testing of moisture content is essential for many species before they are stored for a long time, but is usually unnecessary if the seed is to be sown immediately in the nursery; on the other hand, partial drying may be needed before despatch of seed to a distant forest station, to reduce the risk of deterioration in transit, and any deliberate drying operation must be accompanied by the testing of moisture content.

Laboratory seed tests aim to provide accurate and reproducible guidance, rather than absolute answers or predictions. Viability, germination and vigour tests all produce results that are usually greater than, and at best equal to, how the seed will actually perform in the field. An appreciation of what viability, germination and vigour measure can help maximise the understanding of the planting value or storage potential of seed.

Seed germination tests assess the ability of the seed to produce a healthy plant when placed under favorable environmental conditions. Germination tests are conducted for a prescribed time period under laboratory conditions that assure optimum moisture, temperature and light. Unfortunately, these conditions are seldom encountered in the field, and field emergence may be overestimated by standard germination tests.

The book details procedures of routine seed testing methods and recent approaches in seed testing.

I thank all members of my team who have helped in the preparation of the book. My special thanks go to "Random Publications" who have published the book.

– ***Dr. Mahesh Kumar***

Contents

1

Management of Seed Testing

INTRODUCTION

Tests of the quality and other characteristics of seed need to be made at several stages during its progress from the parent tree to the seed bed. Tests for maturity and soundness of seed before and during collection in the forest and are usually associated with surveys of the abundance of the seed crop; they are intended to ensure that both the quantity and the quality of seed justify the effort and cost of collection. Several tests may be required at the seed processing depot, after extraction and cleaning and before the seeds are despatched to the nurseries or placed in storage. Testing of moisture content is essential for many species before they are stored for a long time, but is usually unnecessary if the seed is to be sown immediately in the nursery; on the other hand, partial drying may be needed before despatch of seed to a distant forest station, to reduce the risk of deterioration in transit, and any deliberate drying operation must be accompanied by the testing of moisture content.

Testing of germination or viability should be repeated at the end of the storage period if this is more than a few months and, in the case of long-term storage for conservation of genetic resources, testing should be done at intervals throughout the storage period.

Efficiency and success in raising plants in the nursery and in their subsequent establishment in forest plantations depend to a great extent on the quality of the seeds used. It follows that foresters, nurserymen, seed dealers and others need accurate estimates of the quality of the seeds in which they deal or which form the basis of their afforestation projects. This is particularly important when seeds are bought and sold, or move between countries in international trade.

The example of estimating seed demand which is shown in Table uses estimates of the number of germinated seeds per kg of Pinus kesiya and Tectona grandis which are averages for the species. In practice, the number varies considerably from year to year and from seed lot to seed lot, so that it is necessary for the forester to have an accurate estimate of germination for every individual seed lot which he receives, if he is to fulfil planting targets while at the same time avoiding the waste of valuable seed caused by sowing too much. Similarly the nurseryman must have a good estimate of germination if he is to adjust sowing rates to produce the optimum spacing of seedlings in the seedbed.

Justice (1972) has defined the following objectives for developing rules for seed testing:

- To provide methods by which the quality of seed samples can be determined accurately;
- To prescribe methods by which seed analysts working in different laboratories in different countries throughout the world can obtain uniform results;
- To relate the laboratory results, insofar as possible, to planting value,
- To complete the tests within the shortest period of time possible, commensurate with the above-mentioned objectives;
- To perform the tests in the most economical manner.

For the practising forester, the most important object of seed testing is the provision of an accurate estimate of the capacity of a given seedlot to produce healthy, vigorous plants suitable for field planting. In the present context, seed "quality" refers to the physiological vigour of the seed rather than its genetic quality.

The essence of good seed testing is the application of reliable standard methods of examination to ensure that uniform and reproducible results are obtained. Standardization has been greatly facilitated through the adoption by a number of countries of the International Rules for Seed Testing formulated by the International Seed Testing Association (ISTA). ISTA was founded in 1921, drew up its first set of Rules in 1931 and made major revisions of the Rules in 1953, 1966 and 1976. In the early stages ISTA's primary concern was for agricultural seeds, but trees and shrubs have gradually assumed greater importance and the 1976 Rules (ISTA 1976) give (firm) "prescriptions" or (more tentative) "suggestions" as to testing methods appropriate for 61 different genera of these, compared with 26 in the 1953 Rules. The Rules are published in English, French and German.

Although tree species and genera from the tropics and southern hemisphere (notably eucalypts) have begun to figure in the ISTA lists, the balance is still overwhelmingly weighted towards north temperate species. Thus important species such as Tectona grandis, Pinus patula, P. oocarpa and P. kesiya are listed, but only in the list of "suggested" (non prescriptive) seed testing methods, while a recent survey (ISTA 1981 a) has shown that 20 other important tropical species are not included at all; they include Cupressus lusitanica, Gmelina arborea, Cordia alliodoraand the genera Albizzia, Araucaria, Casuarina, Swietenia, Terminalia and Triplochiton. Their omission simply reflects the lack of reliable information from research on the best methods of testing these species and genera. ISTA testing methods involve the maintenance of controlled laboratory conditions and the use of some items of equipment which are relatively expensive. They are therefore well suited to large, well-equipped seed laboratories but are impracticable for small forest stations which have no laboratory and must carry out germination tests in the nursery or the office. Lack of equipment is no reason to omit seed testing altogether. For example simple germination tests carried out in the nursery can produce results which are perfectly satisfactory for local use. But precise details of method should always accompany results whenever this is not standardized, so that the seed user can interpret the results in relation to his own nursery conditions.

An up-to-date guide to the range of equipment available for seed testing is contained in the "Survey of Equipment and Supplies" published by ISTA (ISTA 1982). It lists equipment by type, use, model specification and supplier. The earlier directory, "Equipment and supplies for collecting, processing, storing and testing forest tree seed" includes seed testing equipment and gives the addresses of user laboratories as well as suppliers.

Van der Burg *et al.* (1983) give a description of how a seed testing station in tropical or subtropical areas could be established. Two alternatives are described: Seedlab 2000, that can test about 2000 samples per year and Seedlab 5000 that can test at least 5000 samples per year. Directives and general considerations concerning the staffing, the organization of the work, the lay-out of the building, and the equipment needed are given. Forty-six figures and two tables give an impression of the equipment and administrative forms used. Equipment which in the experience of the authors has been found suitable for the work is recommended and detailed descriptions and company addresses are mentioned. Some equipment that is not commercially available is described and plans for construction are included. A list of books and journals for a basic seed testing library is appended.

The tests which may be required are purity, authenticity, seed weight, germination, indirect testing of viability, moisture content, and seed health and damage. A prerequisite for all testing is good sampling, which is described in the following section.

NON-MECHANICAL METHODS OF DIVIDING

HALVING METHOD

The sample is placed on a clean surface and thoroughly mixed by hand; it is then divided into quarters with a sharp-edged spatula and opposite quarters discarded. The process is repeated until a final sample of the approximate weight required is obtained.

Random Cup Method

A series of small cups or thimbles are arranged on a tray in a definite pattern and the sample is poured systematically over this area. The working sample is obtained from randomly selected containers. A modification of this method uses a tray divided into an equal number of square compartments, every alternate one of which has no bottom.

MECHANICAL METHODS OF DIVIDING

Most mechanical dividers are designed to split the sample into two approximately equal parts. The working sample is obtained by repeatedly dividing the bulk sample until it reaches the required weight. ISTA recommends three types of dividers as being suitable equipment. They are the conical divider (Boerner type), the soil divider, and the centrifugal divider. They are briefly described as follows:

Boerner Sampler

This divider is made in different sizes. The essential parts consist of a hopper, inverted cone, and a series of baffles directing the seeds into two spouts. The baffles form alternate channels and spaces of equal width. They are arranged in a circle at their summit and are directed inward and downward, the channels leading to one spout and the spaces to an opposite spout.

SEED SAMPLING TESTING

When seed is tested, the sample on which the test is made has to be representative of the whole. No matter how accurate the technical work in the test, the results can only show the quality of the sample submitted for analysis (Aldhous 1972). Therefore, every effort must be made to ensure that the submitted sample accurately reflects the composition of the whole seed lot. This is equally applicable, whether a whole series of tests is to be made in a laboratory, or a simple nursery test carried out to determine number of germinated seeds per kg of uncleaned seed. Time spent testing carelessly drawn samples may be time wasted.

Completely homogeneous seed lots would be easy to sample, but they do not exist. Common sense measures can be taken to reduce heterogeneity as much as possible, for example seed lots of the same species should not be mixed if they come from different origins (provenances), differ greatly in age or, in the case of the same introduced provenances, if they come from plantations growing on very different site types in the introducing country. But a seed lot may be heterogeneous even if collected from a homogeneous stand. Paul (1972) has given the simple example of pine seed transported over bad roads in the back of a landrover. The bumps and vibrations to which it is subjected will cause a settling out of different types of material. All the small and empty seeds and other light rubbish will eventually reach the surface layers and a sample taken from the top alone will give a completely false impression of the potential performance of the seeds as a whole.

MIXING

If the seed lot is a small one consisting of only a few kg of seed, it is possible to improve its homogeneity by thorough mixing of the whole lot before taking a sample. In the case of very large seed lots transported and stored in many different containers, mixing of the whole seed lot is impracticable.

Instead, a number of different samples are taken, as described on pp. 194–195, and it is these samples which are thoroughly mixed together to form a homogeneous "composite" sample for testing. The methods of mixing which are quoted below from Paul (1972), are equally applicable to small entire seed lots or to composite samples.

"*Methods of mixing. There are two simple ways in which seed may be thoroughly mixed*:

- Mixing with the aid of a mechanical divider. Mechanical dividers are used to reduce the size of seed lots or samples by repetitive halving. Their operation is described below. The divider has the added advantage that it can also be used to mix seed in the following manner:
  - Pass entire seed lot through divider.
  - Take the two equal portions and feed them simultaneously into the divider.
  - Repeat operation No. 2.
  - Repeat operation No. 2 once more.
  - Take the two equal portions and pour them simultaneously into the storage container.
- *Handmixing*: It is surprising how difficult it can be to achieve a homogeneous seed lot using this method. The entire seed lot must be spread out on a sheet of paper or some other suitable smooth surface and mixed by scooping the seed from side to side and from top to bottom. After thoroughly mixing, spread the seed out evenly and divide the lot into four equal parts. Place these separately in four containers and with the aid of an assistant pour them simultaneously into the storage container. Repeat the spreading, quartering and pouring process another two times."

USE OF SEED TRIERS

For a large seed lot contained in a number of different containers, seed triers are used to take small sub samples from different parts of the lot. All of these "primary" samples are mixed together to form a "composite" or bulk sample, which is later reduced in size by successive dividings until it is small enough to form the "working" sample on which the various tests are carried out. As described by Turnbull (1975 d), a seed trier is a probe, long enough to reach all areas of the seed bag and designed to remove an equal volume of seed from each area through which it travels. Most commonly used is a sleeve-type trier which consists of a hollow brass tube inside a closely fitting outer sleeve.

The tube and sleeve have open slots in their walls so that, when the tube is turned or slid, the slots in the tube and the sleeve coincide and seeds can flow into the tube; when the tube is given a half turn or slid in the opposite direction, the openings are closed.

Seed triers should have a series of separate compartments and not just a single long tube with several openings. Tubes are available in varying lengths and diameters to cater for different sizes of container and kinds of seeds. Triers known as "thief" triers, with a single open slot, should be avoided, as they can damage the seeds.

Ideally, the primary samples should be distributed proportionally to the volumes of the different parts of the seed lot. For example, if a lot is contained in ten 10-kg containers and ten 20-kg containers, the 20-kg containers should contribute two thirds of the composite sample and the 10-kg containers only one third.

For seed lots in containers of uniform size, ISTA has detailed prescriptions as to the number of primary samples appropriate for varying numbers of containers, *e.g.*, for 6 – 30 containers at least one in three containers is to be sampled and never less than five.

Large seeds and others that are not free-flowing are not suitable for sampling by seed trier. They should be sampled by thrusting the hand into the seeds and removing small portions. The hand should be inserted flat with the fingers extended together. The fingers should be kept together as the hand is closed and withdrawn. It is difficult to sample with this method deeper than about 40 cm and containers may have to be partially emptied to facilitate sampling.

REDUCING THE SIZE OF COMPOSITE SAMPLES

Usually the composite or bulk sample for testing needs to be reduced to a working sample of standard weight. Whenever possible the sample should be subdivided by a mechanical divider to reduce any operator bias in the dividing.

The non-mechanical halving method is described here for use if a mechanical divider is not available, but it is the least desirable method for most types of seed. Manual methods must, however, be used for seeds which are not free-flowing.

DETERMINING SEED QUANTITIES

Seed users need to define the quantity of seed needed of each species, provenance or stand. For this it is necessary to know the area of plantation to be established annually and the initial spacing to be used, together with an estimate of losses and culls in the nursery, of replacements needed after planting to achieve full stocking, and of the number of germinated seedlings to be expected from each kg of seed sown. An example of the type of calculation is shown in Table.

Information on planting area and initial spacing is usually available from Plantation Management Plans, while some guidance to germination rates is available in published documents. Whenever possible, local experience on variation between provenances and planting sites should be used to refine estimates based on average conditions.

For example, seeds of two provenances of Picea abies weigh, respectively, 6 gm and 12 gm per 1000; seed of Eucalyptus cloeziana collected in the moist coastal forests of Queensland averages 100,000 – 400,000/kg, whereas seed from dry inland wood lands averages only 35,000 – 65,000/kg. In Italy it was found that in nursery trials on several eucalypt species the number of plants produced as a percentage of viable seed ranged from 18 per cent for E. robusta to 46 per cent for E. camaldulensis.

Similarly, differences in climate, soil and incidence of pests and diseases can have a big effect on the rate of losses in different nurseries and plantations, whether or not there are any differences in the efficiency of management. So it may be necessary to apply an appropriate "locality correction factor" or "nursery recovery factor" to arrive at an accurate estimate of seed requirements for a particular plantation project.

Before sending in his final order for seed to a central seed unit or commercial seed merchant, the plantation project manager should deduct those quantities of seed already in stock or likely to be available by collection from older plantations within the project area.

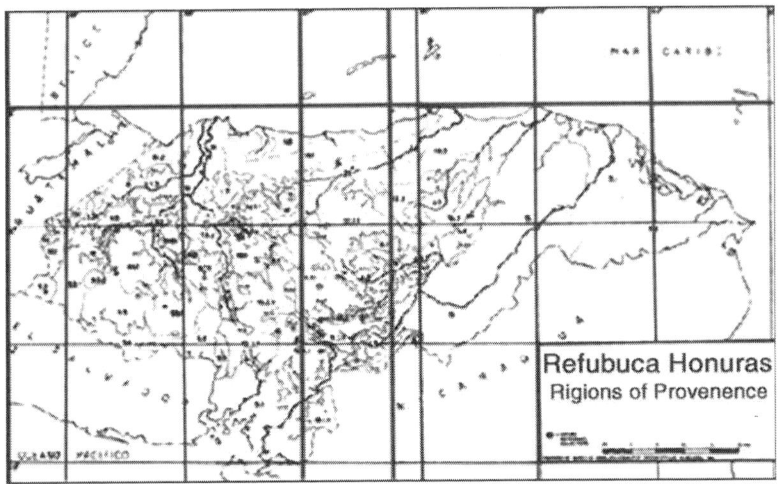

Fig. Provenance Regions for *Pinus caribaea* and *P. oocarpa* in Honduras.

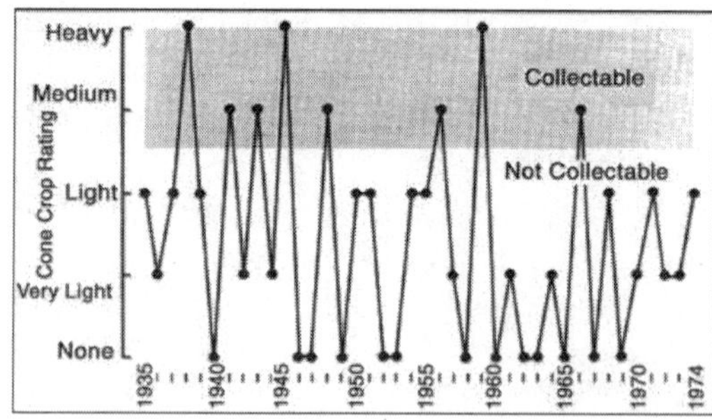

Fig. Cone Crops of Douglas-fir, Vancouver Forest District. Over the 40-Year Period, Only Eight Cone Crops have been Considered "Collectable", *i.e.* Heavy Enough to Justify Mobilizing Large Scale Collections. The Period between Collectable Crops has Ranged from 2 to 8 years.

Fig. Example of a Cone-Cutter for Seed Crop Estimation on a Longitudinal Section.

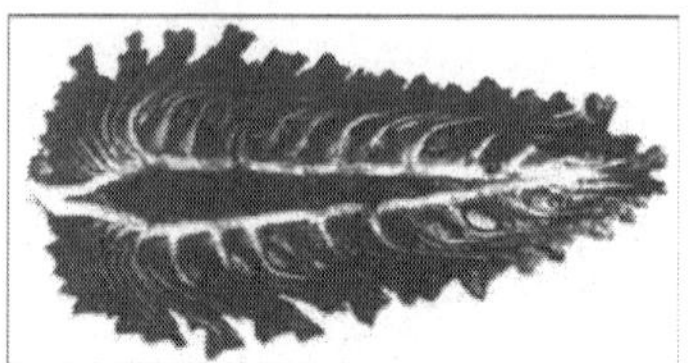

Fig. Seed Content is Estimated by Counting Good Seeds on One Surface of Each of Several Sliced Cones

An alternative approach, favoured in some countries, is for the project manager to specify the number of plantable plants he needs to raise and to leave it to the seed officer to decide on the weight of seed to be collected and issued, in the light of germination tests of current seed lots and of known nursery recovery factors.

If good annual crops of seed can be guaranteed, it is satisfactory to order enough seed each year to produce nursery stock for the area scheduled for planting in about two years' time. Storage space can then be kept to a minimum. For species which show periodicity in seeding, however, there are big advantages in acquiring several years' supply in a single good seed year, when the seed will be both cheaper and better. This course of action can only be of practical value if local seed storage facilities are adequate to maintain seed viability over the interval between good seed years. Knowledge is needed for each species or provenance on both the likely interval between good seed years and the rate of loss of viability of the seed under existing storage conditions. If heavy seed years are separated by several successive years of seed crop failure and if storage at ambient temperature results in rapid loss of viability, the alternatives may be to build a refrigerated storage room or to change the species.

SEED LABORATORY METHODS

DRY WEIGHT

The most generally accepted measure of maturity is the time when the seed has reached its maximum dry weight, a point called physiological maturity. This means that nutrients are no longer flowing into the seed from the mother tree. Maximum freshweight does not indicate physiological maturity because the maturing seed begins losing water while nutrients are still being accumulated and biochemical processes are continuing. Recurrent dry-weight determination of a series of seed samples can be made and the results extrapolated to the remaining crop, but this method is slow and therefore seldom used.

CHEMICAL ANALYSIS

Biochemical changes take place as seeds mature but relatively little is known for most species. Chemical indices of seed maturity have been determined in a few species *e.g.* content of crude fat and protein-nitrogen, which increase five and four times respectively from immaturity to physiological maturity, are the best chemical indices for Fraxinus pennsylvanica. But they have no advantages over an examination of the embryo and the colour change of the fruit, and the extra trouble to perfom the analyses does not appear to be justified. Rediske found that Pseudotsuga seeds were physiologically mature when the content of reducing sugars dropped to 14 mg/g.

X-RAY RADIOGRAPHY

The examination of the development of the embryo and endosperm of sample seeds by means of X-ray radiographs is a quick and relatively straightforward method of assessing seed maturity, provided that suitable facilities and skilled technical staff are available. The technique has been used successfully for Tectona and a number of other tropical species, as well as for temperate species such as Pinus strobus. It has the disadvantage of requiring relatively expensive equipment and relies heavily on the judgment of the seed analyst for reliable results.

MOISTURE CONTENT OF FRUITS

Water loss of maturing cones and fruits occurs in many species and is closely related to the maturity of the seed. Seeds of Picea glauca are considered ripe when moisture content falls below 48 per cent, of Larix decidua at 25–30 per cent and of Pinus sylvestris when it falls to 43–45 per cent (fresh weight basis). However, determination of moisture content by drying in an oven suffers the same disadvantage of slowness as determination of dry weight.

COLLECTION OF IMMATURE SEEDS

It is general practice to collect seeds when they are mature, because they have a higher germinative energy and a greater longevity in storage than immature seeds. An alternative method is to collect fruits prior to ripening and to store them in relatively cool, well-ventilated conditions, which permit after ripening of the seeds within the fruit. It has shown promise on a research scale in a number of species.

There are several reasons for the interest in developing techniques for artificial ripening. They are:

TO EXTEND THE COLLECTION SEASON

The short period between seed maturity and dispersal may place an excessive demand on the availability of seasonal labour and in some areas unfavourable weather conditions in the collection period may aggravate this situation. Lengthening the period available for collection permits better organization of the collections and allows skilled personnel to pick more of the crop.

It can be particularly valuable in research, where a large number of seed lots have to be collected from widely scattered localities. Griffin used the technique in his study of Douglas fir provenances.

TO AVOID DAMAGE TO THE SEED CROP BY INSECTS AND OTHER PESTS

Insects, birds, rodents and other pests frequently damage or destroy seeds and fruits when they have reached maturity. Early collection may be one method of avoiding these losses. Damage and deterioration of seeds is usually worse during the period while they lie on the forest floor than during any other stage in their history; any reduction in this period will improve their subsequent viability and longevity in storage.

TO SALVAGE IMMATURE SEED COLLECTED INADVERTENTLY

Untrained collectors of seeds often begin picking fruits and cones too early in the year before they are fully mature. Artificial ripening provides a method for handling this material.

The development of techniques for after-ripening of immature seeds will require more research, before they can be applied to a wide range of species. However, where a problem of rapid dispersal or of seed pests exists and provided the earliest time for safe collection of immature fruits can be established, such techniques can be very beneficial.

DETERMINING WHICH TREES TO COLLECT FROM

If it can be assumed that the seed collector has received clear directions from the seed user as to which species and provenances and, in some cases, from which stands he is to collect, it is still his responsibility to select the individual trees for collection. Criteria will vary considerably according to whether the collections are large-scale for afforestation projects or small-scale for research purposes.

Identification of species presents no problems in monospecific plantations, but is essential and may be difficult in mixed natural forest, especially where very similar species of the same genus occur mixed together as happens with pines in Mexico and Central America, eucalypts in Australia and dipterocarps in S.E. Asia. Unless identification is certain, it is often advisable to collect herbarium specimens as well as seed.

LARGE-SCALE COLLECTIONS

In large-scale collections emphasis is on collecting as much seed as possible, as quickly and cheaply as possible, rather than on very careful selection of the parent trees. It is, nevertheless, essential to avoid collecting seeds from very poor phenotypes or seeds which prove to be empty or non-viable.

Useful guidelines are listed by Stein et al. (1974), and form the basis of the following:

- Collect seed only from healthy vigorous trees of reasonably good form that are making average or better growth.
- Where possible, collect from mature or nearly mature trees. Overmature trees should be avoided, since seeds from them may be of low viability.
- Avoid isolated trees of naturally cross-pollinating species, since these are likely to be self-pollinated. Seeds are likely to be few, of low viability, and any seedlings produced are frequently weak or malformed.
- Avoid collecting in stands containing numerous poorly formed, excessively limby, off-colour, abnormal or diseased trees.

Often it will be necessary to compromise between seed production and phenotypic appearence. No seed should be collected from excessively coarse-branched, vigorous "wolf" trees, even though they often bear a large crop, while trees of exceptionally good form sometimes bear so little seed that they do not warrant the trouble of collection. The bulk of seed will come from trees which are "average or better than average" in both form and seed production. Although few studies have been made of the reproductive biology of tropical trees, the occurrence of some species at a very low stocking (less than one every km^2) suggests that these must be naturally self-pollinated. Seed collection from such trees is free from disadvantages associated with collections from isolated trees of natural cross-pollinators.

SMALL-SCALE RESEARCH COLLECTIONS

In small-scale collections for research, selection of trees will depend on the precise objective of the planned research. Provenance research is now receiving a good deal of attention in many countries.

Advice of IUFRO on provenance seed collections includes the following recommendations on collections from individual trees (FAO 1969):

- Collect from not worse than dominant and co-dominant trees of average quality, within "normal" rather than "plus" stands. Collections from superior phenotypes, if made, should be kept separately.
- Collect from a minimum of 10 trees, preferably from 25 to 50 in the stand. If the stand is very variable, increase the number of trees. Record the number of trees and the approximate percentage which they form of the stand.
- Seed trees to be at least seed fall distance apart from each other. A distance of 100 m has been adopted for Pseudotsuga. This is to reduce the risk of collecting from half-sib parents. In Australia a minimum distance of twice tree height is used as a practical rule of thumb (Boland *et al.* 1980).
- Individual seed trees to be marked.
- Collect equal numbers of cones, fruits or seeds per tree.
- In normal first stage provenance collections, seed from individual trees may be mixed together. If special studies on individual genotypes are to be done, seed from each tree should be kept separate.

SINGLE TREE COLLECTIONS

Foresters are interested in the variation within populations and provenance as well as the variation between them. In the case of exotics, one approach for the introducing countries is to study between provenance differences under local conditions first, and to investigate variation between individuals in the best provenance, by means of progeny trials, only at a later stage after the provenances locally best adapted have been identified. If progeny trials are the object of seed collection, it is essential to keep seed from individual trees separate at all stages of collection, transport, processing, nursery and field planting. The preservation of the identity of individual trees through the collection and extraction phase often requires considerably more effort than bulking the collections.

If the effort is made, then certain advantages accrue, as listed by Turnbull (1975b):

- It permits biosystematic study of genetic variation both within and between populations. McElwee (1969) states that seed collections mustbe kept separate by individual trees throughout the test from collection to outplanting, as provenance tests are weakened by combining seed within a stand, making it impossible to distinguish between seed source and individual variation. However, in large provenance tests with many individual trees, it may be beyond the resources of the investigator to retain the identity of the parents.
- It is possible to manipulate to equalize the amount of viable seed from each tree in the provenance mix if the seed must be bulked prior to sowing.
- It is not always possible to detect trees bearing hybrid seeds in the field. This is particularly the case with eucalypts. If seedlots are kept separate then, following the raising of small samples from each tree, any showing evidence of hybridization can be eliminated before the main trial is established.

SINGLE CLONE COLLECTIONS

In seed collection in clonal seed orchards, the unit of identity to be kept separate is often the clone rather than the individual ramet. In Zimbabwe the practice of keeping individual clonal seedlots separate has been followed for many years and it is considered that it more than justifies the additional costs and efforts involved over bulked orchard collections.

The advantages can be summarized as follows:

- Maintaining individual clonal identities at all stages from collections through to storage enables one to act with the minimum of delay on most information as it becomes available. The more obvious areas where action can be taken quickly are when rogueing becomes necessary or when susceptibility to pests or diseases becomes apparent, and the undesirable seedlots can be isolated or discarded.
- Seedlots can be made up and supplied to suit specific sites, making use of the most recent information on genotype/environment interaction from progeny test analyses.
- In conjunction with (b) above, and using the test results of each clone, composite seedlots can be made up in such a way as to give equal clonal repre-sentation in the final planting stock, avoiding the dominating effects that some clones frequently have in a bulked seedlot.
- In general, it increases the options available to the user.

COLLECTIONS FOR CONSERVATION

Collections are also made for the purpose of attempting to conserve the gene pool ex situ either as seed in long-term storage or in planted conservation stands. Because exact knowledge of gene frequencies in the indigenous populations is largely lacking, collecting for gene conservation must be based mainly on commonsense.

Similar methods as for provenance collections are likely to be appropriate, with the exceptions that:

- A somewhat larger number of trees per genepool should be sampled. Estimates are in the range or 50 to 100.
- The sample should be a strictly random one and include poorer than average as well as better than average trees, in order to capture as much as possible of the total genetic variation. The only restriction on this principle is the impossibility of sampling trees which are bearing no seed.
- An additional measure to ensure the fullest possible genetic diversity within the seed collected is to collect in a better than average seed year. The better the seed year, the better the representation of male parents contributing pollen as well as of female parents supplying seed.
- The quantity of seed to be collected from each provenance will usually be larger, because the recommended area for a conservation stand, 10 ha, is much greater than the total area of one provenance in a single provenance trial.

ASSEMBLING RESOURCES

One part of planning is the timely assembly of clear information on the nature and magnitude of the seed collection tasks—number of species and provenances, seed quantities, location of stands, best dates to collect etc., as described above. The other part is to select and assemble the resources needed to do the job. Details of the various resources which may be useful are discussed in subsequent chapters. During the planning stage, the leader of collecting operations needs to check the preparations for field work under the following headings:

ORGANIZATION OF COLLECTING TEAMS

Known or estimated output of collecting teams needs to be related to the quantity of seed, number of stands and length of season, in order to determine the required number and size of teams. For example provenance collections of Pinus kesiya totalling 3000 kg of cones from 16 stands can be completed in 30 days by one team in Thailand. If the full season is estimated at 45 days and there is a demand for a total at 9000 kg of cones from the same 16 stands, two teams will be needed to do the job.

If planning can be done sufficiently in advance, it may give an opportunity to train additional climbers if they are needed. It is desirable to have at least one tree-climber on the permanent staff, who can be responsible for looking after climbing equipment and for training new temporary climbers. In the field the climbers should be organized in small teams with a foreman in charge of each. In Honduras a team of 6 pairs (with one climber and one ground assistant or anchorman in each pair) has been found to be a suitable size.

ORGANIZATION OF TRANSPORT

Collecting teams need to cut to a minimum the time spent in moving between one site and the next. Transport must be available where and when it is needed. If necessary, extra vehicles may be temporarily hired. In roadless country, advance arrangements may be needed to employ extra unskilled workers to assist in carrying equipment, tents, etc.

ORGANIZATION OF EQUIPMENT

Choice of equipment will vary greatly according to local conditions. The steeper and less accessible the terrain, the simpler and lighter should be the equipment. Whereas highly mechanised equipment such as tree shakers or mechanised platforms may be appropriate in large seed orchards on flat land, only light portable equipment is practical where natural stands are 4 – 6 hours' walk from the nearest road. Apart from collecting tools, safety clothing, first aid equipment and plenty of bags and sacks should be provided.

ORGANIZATION OF RECORDS

Meticulous recording and labelling is essential to good collecting. Appropriate labels and forms need to be designed well in advance and printed in adequate numbers..

ORGANIZATION OF PERMITS

These are not normally required for forest services collecting in government forest reserves, but may be needed when collecting on private land, in National Parks and special reserves, or in another country.

Even if formal authority is not needed, it is often advisable to inform local communities of proposed operations in advance.

ORGANIZATION OF SEED EXTRACTION

Arrangements for rapid movement of fruits from collection site to extractory may be needed, involving the advance organization of transport. The seed extractory staff must be advised when to expect the fruits. If some prelimi-nary sun-drying of fruits in the forest is planned, polythene sheeting or tarpaulins will be needed.

FIELD METHODS OF SEED

SPECIFIC GRAVITY OF FRUITS

As moisture content of fruits and cones decreases with maturation, so does specific gravity or density, the ratio of unit weight to unit volume, decrease. Unlike moisture content, it is not too difficult to determine approximate specific gravity in the field by flotation in liquids of known specific gravity. Specific gravity indices of maturity have been established for cones of a number of coniferous species, and the cone to be tested is placed in a liquid in which it will float if mature and sink if immature. Various mixtures of kerosene (SG = 0.80), light SAE 20 motor oil (SG = 0.88) and linseed oil (SG = 0.93) have been used to prepare flotation liquids having a designated specific gravity. Tests must be made immediately after cones are picked from a tree. Specific gravity indices have proved reliable for some temperate conifers *e.g.* an S.G. of 0.74 for Picea glauca, but not for several southern hardwoods in the USA.

EXAMINATION OF SEED CONTENTS

Examination of seed contents exposed by cutting open fruits or cones lengthwise can be a reliable and simple method of assessing seed ripeness, provided the operator is experienced. Most embryos and endosperm pass through an immature "milk" stage, followed by a "dough" stage when the tissue becomes more firm. Mature seeds have a firm white endosperm (where present) and a fully developed firm embryo.

COLOUR OF FRUITS OR CONES

Colour changes in fruit or cone provide a simple and, in some species, reliable criterion for judging seed maturity, but the operator must be experienced in the characteristics of the species concerned. In common with the specific gravity method, it involves no destruction of the seeds in the sample examined. Colour changes are usually from the green of the immature fruit or cone to various shades of yellow, brown or grey, and this may be accompanied by hardening of cone scales or of the pericarp of dehiscent or woody fruits. Since the seed normally matures before the fruit, it is advisable in some species to time collection at an earlier rather than a later stage of the colour change. Colour change was found to be the most reliable indicator of maturity for general practice in several southern hardwoods in the USA. It has also given good results in a number of temperate conifers. In Malaysia Tamari found that the best results were obtained by timing collection of Dipterocarp fruits when the wings turned brown but before the fruit itself changed colour.

In Thailand cone colour is used as a guide to optimum time of collection of pines, but differs according to species. In Pinus kesiya collection starts when cones have hardened and the colour is changing from green to brown in proportions of 50: 50. In Pinus merkusii optimum time of collection is reached when the majority of cones are brownish and some have started to open (Granhof 1975). Trials with the Zambales provenance of P. merkusii have shown not only that extraction is a much more lengthy and expensive operation with green than with brown cones but also that the seed extracted has lower germination rate. Experience with P. caribaea in Honduras is similar.

Abscission and shedding of fruits is usually a sign of fruit maturity and it might be assumed that it also indicates a high content of sound, mature seeds. This is not always the case. The first seeds or fruits which fall naturally are often of poor quality, in which case it is advisable to reject them and to postpone collection until the peak and the latter half of the season. In Thailand fruits of Tectona grandis start to be shed in March but observations have shown that the most viable fruits are the last to be shed, so collection starting only in April is recommended. The first fruits of Dipterocarp species that fall upon ripening are usually defective and collecting should be delayed until the greater portion of the fruit has fallen.

SEED TECHNOLOGY

Seed technology is the science dealing with the methods of improving physical and genetical characteristics of seed. It involves such activities as variety development, evolution and release of varieties, seed production, seed processing, seed certification and storage. Thus seed technology is interdisciplinary science which includes broad range of subjects. It also includes seed quality control, seed marketing, seed distribution and research on above aspects of seed.

ROLE OF SEED TECHNOLOGY

The role of Improved Seed Technology is Given Below:

- *Improved Seed is Carrier of New Technology:* Introduction of quality seed of new varieties wisely combined with other inputs significantly increases the yield level. In India cultivation of high yielding introduced dwarf wheat varieties helped to increases the production from 12 million tones making country self sufficient with in span of 10 years. Improved Seed is Basic Tool for Secured Food Supply: Due to successful implementation of high yielding variety programme in India crop productivity is significantly increased as a result import is substantially reduced. The export is increasing in majority of crops in spite of rapid population increase. Improved Seed is a Principal Means to Secure Crop Yields in Less Favourable Areas of Production:Good quality seed of suitable varieties helped in securing higher crop yields. In disaster region E.g M-35-1 drought resistant variety in Jawar served as a boon in drought prone area in rainy season. Improved Seed is a Medium for Rapid Rehabilitation of Agriculture in case of Natural Disaster: In the wide spread droughts and floods, it is much economical if government had National Seed Reserve Stock (NSRS) at their disposal. NSRS plays two fold roles. i) It provide improved seed in emergency period in production areas, and ii) To supply seed for resowing in disaster region (Normally which is not available with farmer in time).

GOALS OR OBJECTS OF SEED TECHNOLOGY

Major goal of seed technology is to increase agricultural production through spread of good quality seed of high yielding varieties developed by the plant breeders, well before the planting season at reasonable cost. Following are important objectives of seed technology.

Rapid Multiplication

It is the quickest possible spread of new high yielding varieties and hybrids developed by the plant breeder.

Timely Supply

The improved seed of available well in time, so that the planting schedule is not disturbed and they are able to harvest the good crop.

Assured High Quality Seed

The expected divided in agriculture from use of improved seed is possible with use of good quality seed.

Reasonable Price

The cost of quality seed must be within reach of average farmer. The cost of the seed must be cheap and farmers should able to purchase it.

CONSISTED OF DRYING THE SEEDS

To this date, the most commonly used procedure has consisted of drying the seeds with different procedures until reaching a moisture content of approximately 5–7 per cent. After this, the seeds are placed in containers whose vapour tightness is most often untested and the success is mainly trusted to low temperatures. Ultradrying has been very rarely used, perhaps by no more than 50 banks – mostly of wild species in botanical gardens—out of more than 1500 existing in the world.

The UPM's results have proved beyond doubt the efficiency of ultradrying in orthodox seeds at least under anaerobic conditions and also prove the comparatively lesser importance of temperature for ultradried seeds. It should be noted that moisture content of 4–5 per cent may be enough for the efficient conservation of some orthodox seeds such as those of many legumes (Ellis & al. 1988). However, as this author points out, these seeds do not suffer when ultradesiccated to 1–3 per cent. In a genebank, with thousands of accessions being handled, the need to individualise the methods should be avoided. In this light, a general benefit to the use of ultradrying can be attributed–always speaking of orthodox seeds.

ULTRADRYING WITH SILICA GEL

Ultradrying could be achieved following different ways, but perhaps the most practical one might consist of storing the seeds in balance with dehydrated silica gel. Leaving some gel afterwards with the stored seeds inside the tight container is most useful because the coloured indicator would show any accidental entry of moisture.

Advantages of the use of Silica Gel:

- It provides a practical method to desiccate the seed samples.
- It reaches moisture levels of 2-3 per cent, lower than those obtained by most.
- Tkeeps these levels indefinitely in tight containers.
- It warns over possible anomalies (*i.e.* moisture intake) in the container—when used in combination with a coloured indicator.
- It can be regenerated after its use through a new dehydration process with heat.
- It delays also ageing by absorbing toxic gases produced during the process.

Summary of the advantages of using silica gel with an indicator. Besides those related to ultradrying itself, it is worth noting the last one (number 6), which refers to a totally independent protection mechanism. In seed samples stored within a closed container, damaging gases are generated during the aging process and as they accumulate, they prove deathly for the entire sample. The mortality, which might have been supposed a priori probabilistic for every individual seed, becomes accelerated at the end. Being capable of absorbing these gases, silica gel significantly delays aging and postpones the moment of death. Ethylene is an important component of this mixture of gases (Lee, & al. 2001).

Silica gel is simply silicic anhydride, SiO_2, though amorphous (non crystalline) obtained through an industrial process. It is granular in texture, white and very porous. It is this last characteristic which gives its absorbent properties. When it is well dehydrated and placed on a scale, it can be observed to absorb up to 20 per cent of its own weight of water. If placed in a closed container such as a Kilner jar, it balances itself with the confined atmosphere until this reaches approximately 10–12 per cent of relative humidity.

The name "gel" is here improperly used. We understand by gel some substance of a colloidal nature, which is not this case. However, the use of this term for this product has long been used and is now strongly established.

The colour usually shown by silica gel is due to an indicator, added to see directly when it is dehydrated and when it has absorbed moisture. For many years, cobalt chloride (Cl_2CO) has been used. This substance gives the dehydrated gel a strong blue colour and a pale pink colour to the gel having absorbed moisture. Recently, the European Union banned its use because of considering it carcinogen through inhalation. A search for new alternatives led to some iron salts, where the change in colour can be poorly distinguished. At present, the most advisable alternative is methyl violet, which gives the dehydrated gel an orange colour and a green colour to the hydrated gel.

The gel is manufactured in granules of different sizes. This size is not irrelevant because, to be used in Kilner jars, a gel with a size > 4 mm can be more convenient, while a 2-3 mm size can be better for glass vials. By hydrating with water coarse silica gel granules (be careful, heat is produced!) the grains get fragmented, and it is possible to screen it in fractions with smaller granule sizes. Hydrated silica gel can be regenerated through heating. When it is accompanied with cobalt or iron salts, this is done at approximately 220 °C, thus a standard cooking oven can be enough. The process may last one or two hours - depending on the mass to be regenerated. With an organic indicator (*e.g.* methyl violet) a temperature of 110–120 °C must not been surpassed and the process will require comparatively more time.

FREEZE DRYING

Freeze Drying ("liophilisation") has been successfully used to obtain ultradry levels of moisture. This procedure consists in freezing the seed water and sublimating the ice afterwards. This technique was developed in the National Centre of Natural History (Paris) in the 1980s for long term storage of pollen grains (Schoenke & Bey, 1981. The Conservatoire Botanique National Méditerranéen (CBNM) started using it for seeds in Porquerolles in 1987.

It is assured that freeze dried seeds do not suffer any decrease in their viability because of the process. However, it does not seem to be the case for every species and the details when applying this technique must be optimized for every species (Anonymous, 2006). As it will be explained, freeze drying may have the advantage of achieving a direct ultradrying instead of one done in stages – as is generally the case with silica gel which needs to be regenerated every time it is used. In comparison, Freeze drying needs a rather expensive machine (Hu, X. & al. 1998). In turn, silica gel reaches lower water contents. Also, considering that silica gel is going to play an essential role afterwards in keeping a low moisture and monitoring the samples, it may be more practical a direct drying with gel. In other words, while freeze drying only ultradries, silica gel not only ultradries but also provides an efficient monitoring during subsequent storage.

THE LOW TEMPERATURE FACTOR

The use of low temperatures has been widespread in long term seed preservation. Standards set by the Food and Agriculture Organization of the United Nations (FAO) and the International Plant Genetic Resources Institute

(IPGRI, now Biodiversity) advise a storage temperature of -18 0C or lower (FAO/IPGRI, 1994) and almost everybody now planning a genebank instinctively thinks of a cold room capable of at least reaching -20°C. Localisation of the new installations of the Nordic Seed Bank in a region of "permafrost" responds to the same idea.

Unfortunately, emphasis on low temperatures and neglecting low moisture (either because of not drying enough or because of the use of inappropriate containers) has brought that a significant number of the actual genebanks are worried with the burdensome need to rejuvenate their prematurely aged material.

Seeds which had been freeze dried in the Conservatoire Botanique National Mediterranéen (CBNM) of Porquerolles and then kept at room temperature for 10 years showed a slightly better behaviour than those kept under low temperature. In a much longer term, ultradry seeds with silica gel kept for 40 years at room temperature inside a closet (Pérez-García & al., 2007) did not behave very differently from those kept in the cold room. It appears hence clear that the main role attributed to low temperature during decades should now be replaced by a much more thorough attention to the moisture factor.

However, it is obvious that low temperature helps, and it would be irresponsible not to take it into consideration. In the last mentioned work, it is suggested that temperatures between –5 °C and +5 °C could very well be enough for orthodox ultradry seeds, thus saving considerable amounts of energy.

2

The Seed Security

The seed consists of three basic parts:

- Embryo,
- Storage tissues
- Seed coat.

The embryo has a reproductive function, being capable of initiating cellular divisions and growth. It is the most important part of the seed. It is an axis that originates growth in two directions, with the objective of originating a root and a shoot. Usually, the embryo is very small, compared to other parts of the seed. It is like a miniature plant. It is consists of a radicle, plumule, one or two cotyledons, and hypocotyl or epicotyl, depending on the type of plant. Energy storage is localized in the cotyledons, in the endosperm or the perisperm. Cotyledons originate from the zygote and are part of the embryo. In many species, storage substances are localized in the cotyledons, and the embryo develops absorbing all the endosperm.

Episperm is the seed cover, consists of two layers, the testa and the endopleura. The outer layer is the testa; it can be stony, leathery, membranous or fleshy. Over the testa we can recognize: the hilum, scar or point of attachment of the seed to the funiculus, water penetrates easily through it; Micropyle, the point upon the seed at which was the orifice of the ovule through which the pollen tube enters; raphe, suture originated by the part of the funiculus that is fused along the side of the ovule. The endopleura is the inner layer; it is thin and generally whitish. Teguments, testa or protective covers delimit the seed. They are formed by one or more layers of cells originated from the ovule integuments and sometimes from the pericarp made from the walls of the ovary.

Regarding the healthiness or pathology of the seed we will consider three aspects that allow us to understand processes associated to seed-borne diseases: infection of the seed, infection of the plant and steps that can be taken to reduce the damage caused by this relationship.

INTRODUCTION TO SEED

Seed is the starting point of agriculture and dictates ultimate productivity of other inputs. Good quality seed alone increases the yield by 15-20 per cent. To meet the potential challenge of catering to the food need of 1.4 billion people of our country by 2025, a quantum increase in agricultural productivity is very much essential and hence production and distribution of high quality seeds of improved varieties/ hybrids to the fanning community is becoming increasingly important. The expansion of agriculture under tropical conditions due to the improvement of cultivars with juvenile period imposed a scientific and technological challenges concerning the seed production under different environmental conditions.

The seed programme includes the participation of state government, SAU system, public sector, and cooperative and private sector institutions. With the best efforts of all these organized sectors, only 15-20 per cent of the total requirement of quality seed is being met with. In most, kind of seeds, the farmers depend on their own farm saved seeds for crop production, which needs certain basic practices of selection of good seeds for sowing. Moreover the crops are raised for market and a small portion of the grains are separated, stored and used as seeds in the next season which may not meet the quality aspects as expected for a seed which results in poor field stand, and ultimately yield.

Quality seed is the key input for realizing potential productivity. As the quality deteriorates during subsequent generations the old seed must be replaced with fresh lots of quality seeds. Ideally seed should be replaced every year for hybrids and every three to four years for non-hybrids. Therefore it is necessary to improve the availability of quality seeds to raise the Seed Replacement Rate (SRR). Despite implementation of the organized seed programme, the seed replacement rate has only reached 15 per cent and there exists an alarming gap between the demand and supply of quality seeds.

The role of private seed industry in the production and distribution of quality seeds is well recognized in the Indian' seed industry. However, they remain generally in the production of low volume high value seeds, which cater. to the needs of only few selected farmers..

The distribution of high volume low value seeds such as rice varieties, oil seeds and pulses are still with the public sector organization. The non-availability of quality seeds in oilseeds and pulses is one of the main reasons for its lower seed replacement rate.

The immediate increase in the productivity and production of these corps can be achieved by a higher distribution of quality seeds of new and high yielding varieties. There is vast scope to produce and distribute quality seeds in these corps for which seed village concept is a noval and highly practical approach and needs to be promoted to facilitate production and timely distribution of quality seeds of desired varieties at village level. In this context, the concept of seed village which advocates village self-sufficiency in production and distribution of quality seeds is getting momentum.

Seed technology is the science dealing with the methods of improving physical and genetical characteristics of seed. The various aspects coming under seed technology are seed production, seed processing, seed certification, seed testing, seed storage, seed biology, seed entomology, seed pathology and seed marketing.

CONCEPT

- Organizing seed production in cluster (or) compact area
- Replacing existing local varieties with new high yielding varieties.
- Increasing the seed production
- To meet the local demand, timely supply and reasonable cost
- Self sufficiency and self reliance of the village
- Increasing the seed replacement rate

FEATURES

- Seed is available at the door steps of farms at an appropriate time
- Seed availability at affordable cost even lesser than market price
- Increased confidence among the farmers about the quality because of known source of production
- Producer and consumer are mutually benefited
- Facilitates fast spread of new cultivars of different kinds

INFORMATION

In many countries of the world the number of trees planted increases each year. A recent survey (Lanly 1982) has estimated that the area of forest plantations in tropical countries will increase from 11.5 to 17.0 million ha between 1980 and 1985, an increase of 48 per cent in five years. Of the 11.5 million hectares at the end of 1980, 40 per cent had been planted in the previous five years.

The average annual planting rate is expected to increase from 0.92 million ha in the period 1976 – 80 to 1.10 million ha in the period 1981–1985, an increase of 20 per cent. Large planting programmes are also in operation in many temperate countries. In addition to new planting, replanting of harvested plantations of non-coppicing species must be carried out on a considerable scale each year. Large as these areas may appear, they amount to only about a tenth of the area of natural forest being destroyed in the tropics over the same period. A further increase in planting rates beyond 1985 will certainly be necessary.

Forest plantations are a powerful tool in the continuing efforts of foresters to increase productivity per unit area—the only means of reconciling the increasing demands for forest products and services on the one hand with a decreasing area of land available for forestry on the other. A combination of intensive site preparation with the use of uniform, well-grown nursery stock, planted at uniform spacing, increases growth and yield, reduces rotation length, facilitates tending and harvesting operations and improves the quality and uniformity of wood, as compared with natural forest.

Plantations also offer the means of using on a large scale the genetically improved material developed by tree breeders. Although there is no case for the indiscriminate replacement of all natural forest by plantations, their judicious use, by providing an alternative source of forest products, can itself reduce pressure on the remaining natural forest and so help to conserve it as a habitat and a source of genetic diversity. Not only do plantations have a major role as producers of timber, pulpwood and wood-based panels for forest industries, but fuel wood and pole plantations and farm woodlots are locally important in many countries. Tree-planting is not confined to block plantations. Shelterbelts and dispersed planting for soil stabilization, habitat improvement, urban and rural amenity or as part of an agrisilvicultural system all benefit the human environment.

With such a variety in planting purposes, it is not surprising that the scale of tree-planting and the variety of species planted continue to grow in so many countries. The current greatly increased interest in agro forestry opens up a whole new range of species for trial. Ability to grow in symbiotic relationship with agricultural crops will be the essential characteristic and will involve criteria such as rooting habit, ability to fix nitrogen and multipurpose uses (food, wood, shelter). Low stature may be beneficial and shrubs may become as important as trees. These new developments will introduce new opportunities and new problems in seed collection and handling. With a few exceptions, notably among the poplars and willows and in some tropical species of Casuarina, trees are propagated from seed, and the suitability and quality of the seeds have a big effect on the success of the plantations raised from them. The use of sound seed from stands of high inherent quality is widely recognized as the best means of ensuring fast-grown and healthy plantations capable of yielding high quality wood. Seed quality comprises both genetic and physiological quality. It must be stressed that "good seed" implies seed which is both of high viability and vigour and is genetically well suited to the site and to the purpose for which it is planted. Physiologically good seed may lead to successful establishment of a plantation but this is of little value if it is slow growing, ill adapted to the site or produces the wrong kind of wood because the provenance or genotype was incorrectly chosen.

On the other hand, there is little point in producing genetically improved seed at an increased cost if it is killed by poor handling techniques and has to be replaced or supplemented by inferior seed in order to achieve planting targets. Good seed handling is an essential complement to genetic improvement. Quantity, as well as quality, of seed production is important. In natural stands variation in the quantity of seed produced affects the forester's decision as to the years in which to collect seed and the trees from which to collect it, as described in Chapter. More intensive management affords him the opportunity of stimulating heavy seed production in genetically superior crops through deliberate treatment, *e.g.* thinning in seed stands or a combination of initial spacing, irrigation, fertilization and thinning in seed orchards. Description of these methods is outside the scope of this book, since they are properly a part of seed stand and seed orchard management, but programmes of seed supply should include measures to improve production quantitatively, as well as improving the genetic quality of the seed and the efficiency of collection and handling.

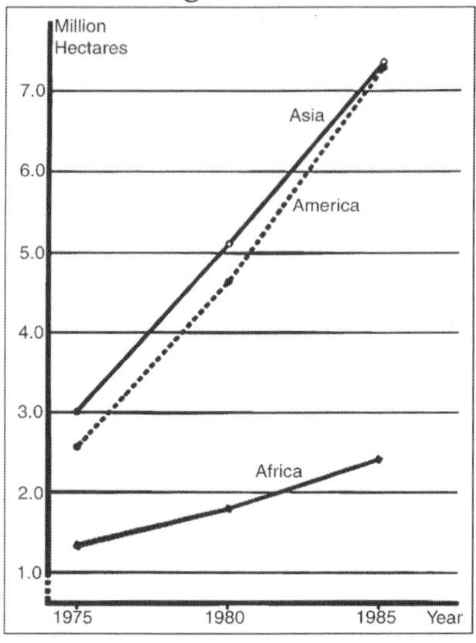

Fig. Estimated Total Areas of Forest Plantations in Tropical Countries

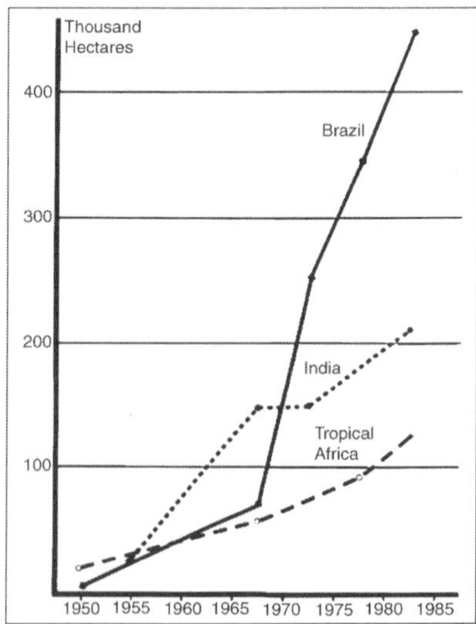

Fig. Estimated Annual Planting Rates
Industrial and Non-industrial Forest Plantations.

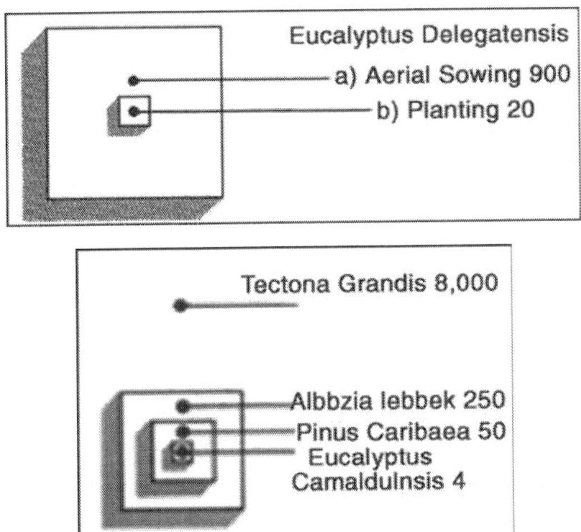

QUANTITY OF SEED

There is a considerable body of published information on seed handling in the temperate zones but published accounts of tropical experience with forest seeds are scattered and incomplete. While recapitulating the principles of seed handling derived from temperate experience, this guide seeks to illustrate these, as far as possible, by examples taken from tropical species. At the same time it must be emphasized that there is great variation in seed biology and that certain techniques which are commonly practised with good results in the temperate zones, *e.g.* stratification or prechilling, may have no useful function whatever in the tropics. It is dangerous to extrapolate temperate experience to tropical species, experience in the dry tropics to rain forest species, or experience in tropical agriculture to tropical forest tree seeds, without testing it in each case.

Valuable lists of equipment and suppliers have been published in the last decade. That published by ISTA (1982) covers equipment used in seed testing. The older list compiled by Bonner (1977) covers all operations from seed collection to seed testing and includes the names of equipment users as well as suppliers. Both equipment and techniques must be adapted to local species and conditions. For example hard-coated seeds of high natural longevity which are produced regularly every year require no more for storage than a well-ventilated room with protection against pests and diseases. A domestic refrigerator or "deep freeze" may be perfectly adequate for long-term storage if only small quantities are to be used each year. Large units and mechanized equipment may be essential in some countries, but heavy expenditure on such items should be incurred only after a full evaluation of the alternatives.

The chain of seed handling operations is only as strong as its weakest link. Therefore great care must be taken at all stages if seed viability is to be maintained all the way from the parent tree to the nursery bed. If a seed has lost viability in an early stage of the process, the best storage or pretreatment can never bring it back to life. Perfect extraction and cleaning of seeds are a waste of money if they are killed later by incorrect storage conditions or careless handling in transit. The greatest risks to seeds occur during temporary storage immediately after collection, during transit to the processing depot, and again during transit from the seed store to the nursery. These are the periods when sophisticated equipment is not available and everything depends on common sense and meticulous care to ensure good ventilation and avoid extremes of temperature.

"International exchange of tree seeds is an opportunity to share the world's forest wealth" (Baldwin 1955). The widespread success of species such as Pseudotsuga menziesii, Pinus radiata, Eucalyptus globulus and Tectona grandis is a convincing demonstration of the advantages which many countries can derive from the introduction of exotics. International cooperation in seed procurement is now practised by many countries and has long been supported by international organizations such as FAO and IUFRO. It should be increased. At the same time the movement of seed between countries does involve some additional problems in maintaining both the viability and the identity of the seeds. Phytosanitary regulations, including inspection and treatment, must be complied with in order to reduce to a minimum the risk of importing new and dangerous diseases. But every effort must be made to eliminate all unnecessary delays in transit, including those imposed by customs, plant hygiene and air lines, and to avoid excessive or duplicated fumigation treatments which kill the seeds. Both sender and recipient must be aware of all the current regulations in force in the exporting and importing country and must plan the seed shipment well in advance to ensure smooth and unimpeded movement of the consignment.

Apart from a few well-known species such as Tectona grandis, research on tropical forest seeds has been inadequate in comparison with both the severity of the problems and the large number of species of potential value for plantations. Much has still to be learned. A first step to this is a good understanding of the natural reproductive biology of each species. For species of the dry tropics which survive naturally by means of seedcoat dormancy, storage should present little problem and a modest programme of research should indicate the most appropriate pretreatment to overcome dormancy and induce uniform nursery germination. The problem of preserving viability in recalcitrant seeds of tropical rain forest species, especially those which will not survive temperatures below about 10° C, is a much more intractable one. A variety of possibilities has been suggested (King and Roberts 1979) but so far little progress has been made towards a practical method of storage which would be applicable in large-scale afforestation projects. Much more research is needed to solve this problem. Until a solution is found, the dipterocarps and other recalcitrant species of the tropical rain forests will remain as much non-starters for widespread plantation forestry as they have been in the past.

SEED INFECTION MECHANISMS

The area of science that studies the relationship between pathogens and seeds is Seed Pathology. It does not only identify the pathogens, it also includes the role of the seed as source of inoculum, the survival of the pathogen and the actions taken to control the pathogens associated to it. It uses the knowledge of General Pathology, Microbiology and Seed Analysis. To obtain a perspective of seed borne diseases, seed borne microorganisms can be considered fewer than four classes.

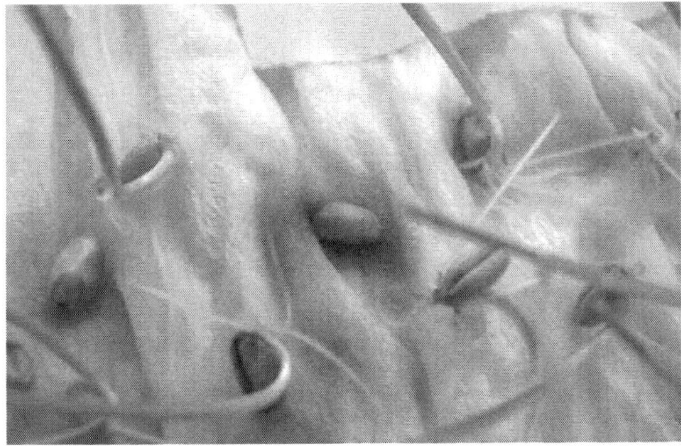

Fig. Infection of the Seed

The first consists of pathogens for which the seed is the main source of inoculum; when seed infection is controlled, the disease is controlled. An example would be lettuce mosaic virus. For these pathogens, the importance of seed-borne inoculum has long been recognized, and control practices have been developed. The second class consists of important pathogens in which the seed borne phase of the disease is of minor significance as a source of inoculum.

Examples are those in which the crop residues in the field were the major source of inoculum. The third and largest group of seed borne microorganisms consists of those that have never been shown to cause disease as a result of their presence on seeds. The fourth class is a group of microorganisms that can infect the seed either in the field or in storage and reduce yield and seed quality. Examples of field fungi are *Diplodia, Fusarium, Cladosporium,* etc.

The storage fungi *Aspergillus* and *Penicillium* can invade most types of seeds under high-moisture storage conditions. The process of seed infection is influenced by the conditions under which the crop grows. Between the facts that influence in the process of infection are: the host and its genotype, the pathogen and its genotype end environmental facts. An infected seed will not always an infected seed will be the cause of the infection of the infection in the plant which originates, so, it is to say that the infection caused from an infected seed is an exception, not something to happen generally. There are two situations, Systemic infection of the seed and Contamination or Infestation of the seed.

Systemic Infection of the Seed

The establishment of a pathogen in any part of the seed is refereed as seed infection. It can be systemic, by the vascular system or plasmodesmata or directly by natural or artificial wounds. The same pathogen can infect the seed using one or more of these mechanisms. For example: *Xanthomonas campestris* pv. *phaseoli* can infect seed through the vascular system, by natural openings (From the pod suture goes to the funiculus, then to the raphe and tegument, or it can also happen thought the micropyle).

Systemic Infection through Flowers, Fruits or Funiculus

Most of the systemic seed-borne bacteria and fungus reach and infect the embryo through the flower or from the peduncule of the fruit, via funiculus. Viruses go to the embryo from the systemically infected mother plant and the infected or contaminated pollen. They rarely reach the embryo during the formation of the seed or the embryo itself.

Examples of some infections that occur through the vascular system are: Some forma specialis of Fusarium oxysporum in pumpkin, pea and tomato; Plasmopara halstedii in sunflower; Septoria glycines in soy; Verticilium dahliae in spinach and sugar beet; Certain pathovares of Xanthomonas campestris in bean, cabbage, rice and sweet pepper; and Pseudomonas syringae pv. Lachymans in cucumber.

Penetration through the Stigma

During the infection, some pathogens follow the same path as the pollen grains do. Spores of some fungus reach the stigma and germinate, producing an hypha that reaches the ovary through the style, where they can stay as a dormant mycelium until seed germination.

For example: Ustilago nuda and U. triticci in barley and wheat, and Alternaria alternata in sweet pepper. Viruses can infect through infected pollen, the male gamete carries the virus, when joining the ovule it generates an infected embryo. If both, the male and female gamete are infected they can even produce an infected endosperm.

Penetration through the Wall of the Ovary

Some fungi, like *Ustilago nuda* and *U. triticci* penetrate through the wall of the ovary as a result of the germination of the Teliospores on the stigma or the wall of the ovary. The pro-mycelium goes through the wall and other tissues until it reaches the embryo. In some other cases, penetration occurs through breakages on the testa, establishing itself in the endopleura or the endosperm. In fleshy fruits, like cucumber, melon, eggplant, tomato, sweet pepper and others, contamination can occur directly through the funiculus or in the tegument, during the process of seed formation.

Examples of this are Coiletotrichum lagenarium in watermelon and Rhizoctonia solani, when it invades fleshy fruits, it is capable of infecting from the placenta and penetrate to the developing ovule or seeds that are still in its formation process and have not lignified its cover.

Penetration through Wounds and Natural Openings

Natural openings like the hilum and the micropyle or wounds generated during the thresh are spots where pathogens like *Xanthomonas campestris* pv. *phaseolicola* in bean and *Pseudomonas syringae* pv. *lachrymans* in cucumber.

SEED CONTAMINATION OR INFESTATION

Contamination or infestation refers to the passive relationship of a pathogens and seeds. The pathogen itself or parts of it can stick or can get mixed with the seeds during any of the processes during seed recollection: harvesting, extraction, thresh, selection and packing.

Pathogens that Stick to the Surface of the Seed

Pathogens that stick to seeds during harvest or postharvest do so by their spores (Clamidospores, Oospores, Teliospores, Uredospores), bacterial cells and in some cases, virions. Spores of the following fungi can be carried on seed coat surfaces: Alternaria brassicae and A. brassicicola in crucifers; A. longipes in tobacco; A. radicina in carrot; Ascochyta pinodella in pea; Drechslera sorokiniana and D. oryzae in rice; D. avenae in oats; Fusarium oxysporum f. sp. callistephi in China aster; Sclerotia of Rhizoctonia solani in eggplant, pepper, and tomato; Tilletia caries, T. foetida, T. contraversa, and Urocystis agropyri in wheat.

A number of bacteria, such as the following, contaminate seed surfaces: Corynebacterium flaccumfacjens pv. flaccumfacjens in bean, Pseudomonas syringae pv. phaseolicola in bean, P. syringae pv. lachrymans in cucumber, P. syringae pv. tomato in tomato, C. rnichiganense pv. michiganense in pepper, *Xanthomonas campestris* pv. *campestris* in cabbage Also some viruses as Tobacco mosaic virus, Tomato mosaic virus, Pepper Mosaic virus.

ACCOMPANYING CONTAMINATION

This type of contamination refers to physical mixing of the seed with pathogen's propagation organs like the sclerotium, nematode's galls, contaminated plant parts or soil particles containing pathogens.

Structures of the Pathogens

Some fungi produce resistance structures named sclerotiums that are made of compacted mycelium. Under certain conditions of temperature and humidity it can germinate, it is common that this occurs together with the seed, producing its infection, by direct penetration or by the spores generated by the fructification organs (Apothecium or Perithecium). These spores are carried by the wind and taken to susceptible tissues of the plant, generally, some parts of the flower. Sclerotiums have many shapes and during thresh can be easily included with the rest of the seed.

Some examples of this are Sclerotinia sclerotiorum in horticultural, grain and flower crops, Sclerotinia cepivorum in garlic and Claviceps purpurea in barley.

Mix with infected Plant Parts

Infected plant's parts and residues can carry fructifications or spores of fungi and bacteria. This situation is particularly common in cereal, forage and oily crop's seeds. Examples of this are Septoria nodorum in wheat, Puccinia malvacearum in hollyhock (malvarrosa), Colletotrichum trifolii in clover, Erwinia carotovora pv. carotovora in tobacco, Pseudomonas syringae pv. phaseolicola in bean and Sclerotinia sclerotiorum in sunflower, soy and peanut.

Soil

Seeds can be mixed with contaminated soil and, carry micro-sclerotiums of *Macrophomina phaseolina* in dirty seeds of kidney bean, Verticillium oxysporum f.sp. phaseoli in cotton seeds, Plasmodiophora brassicae in turnip and Fusarium oxysporum f. sp. phaseoli in bean.

SEED TRANSMITION

We saw how the seed gets infected or infested. Now we will see how these seeds can produce an ill plant. This is to say, how the inoculum goes from the seed to the plant. We call seed-borne pathogens only those that can produce an infection. Because they have to be distinguished from those that can associate to the seed but do not produce an infection, these ones are called pathogens not transmitted by seed.

The virus that causes curly top in sugar beet can be present in the perisperm but does not cause an infection in the plant. In general terms, infection can be classified into systemic and non systemic. It is systemic when the pathogen introduces itself to the plant when the seed germinates, and develops with it. Non systemic infection occurs when there is a localized infection caused by the pathogen in the seedling at the stage of pre or post emergency, in this situation there are no systemic symptoms.

Systemic Transmition

This type of infection can be produced by pathogens that are carried with the seed in various parts, like the embryo, endosperm or episperm, or by contamination of the outer portion of this one.

Infection of the Embryo

When the seed germinates, if the embryo is infected, the pathogen initiates its growth together with the plant. Symptoms can show up during different stages of development. In the case of *Ustilago triticci,* the mycelium grows together with the plant and expresses symptoms only at the flowering stage, which are that all the tissues of the spike, exept the rachis, are replaced by spores.

Some pathovares of *Xanthomonas campestris* that infect cabbage, bean or sweet pepper move between the cells of the host until they reach the vascular system and produce symptoms in leaves or stems. Most of seed-borne viruses persist inside the embryo. Its multiplication and movement accompanies the plant during its development, and there can be symptoms at any stage, from the formation of first leaves until flowering or fructification.

Non Embryo Infection

Infection of the episperm (testa and endopleura) occasionally conducts to a systemic infection. Some bacteria, like Corynebacterium michiganense pv. michiganense in tomato and Xanthomonas campestris pv. campestris in cabbage, penetrate through stomata of cotyledons, and from there, reach the vascular system, initiating the systemic infection.

Episperm Contamination

In some few cases this type of contamination conducts to a systemic infection. Most of these exceptions are fungi that are highly specialized in their pathogenesis and produce in cereals the so called smuts, rusts or mildews.

In these cases, generally, spores are carried outside the seed, they germinate, penetrate the coleoptile, and start a systemic infection. This can occur directly o by a more complex system of haploid hypha fusion in genera like Tilletia, Ustilago, Uromyces, Sclerospora, Pernospora and Puccinia.

Non Systemic Transmition

Non systemic infection is very common, and in the same way as systemic infection, it can come from an infection, outside contamination or by pathogens mixed with the seeds.

Infection of the Embryo

This case is restricted to some pathogenic fungi that maintain itself in the embryo or the episperm as hypha inside the seed. Primary infection starts as injuries in the cotyledons or primary leaves, stems or petiole. Fungi fructifications (Pycnidium, Acervulus) can develop on these organs, under certain favorable conditions (temperature and humidity).

These fructifications produce spores that, with the action of water and wind, disperse the disease to other parts of the plant and other plants. This occurs in *Ascochyta pisi* in pea, *Colletotrichum lindemuthianum* in bean and *C. truncatum* in soy.

Infection of the Episperm

Generally, seeds in which the episperm is infected, do not geminate, or germinate and contaminate the soil. Rarely produce a systemic infection, but can infect the seedling from the outside. Over the injuries, new inoculum is produced; this one infects other parts of the plant or other plants. *Septoria nodorum* in wheat under high humidity conditions forms Picnidium in the coleoptile, whose spores disseminate the disease to other plants.

Contamination of the Episperm

Contamination outside the seed's testa can produce healthy seedlings, but the inoculum infects the soil and from there it can cause infections at more developed stages of the plant. Wheat and rice grains that are contaminated with Teliospores of Neovossia indica in wheat and N. torrida in rice, produce Esporidisporas in the soil, that can be carried by the wind and infect flowers, originating the smuts.

Accompanying Contamination

The mixture of seeds with sclerotiums of the fungi or contaminated soil particles produces non systemic infections, at any stage of the plant growth and development. It is common for *Sclerotinia sclerotiorum* to produce mycelium from the sclerotiums. These mycelium can directly infect the plant or produce fructifications that produce spores that can be carried by the wind and infect flowers of different species, like sunflower, soy, peanut, etc.

Prevention of Seedborne Diseases

Prevention of seed-borne diseases is based in two main facts. On one hand there is the usage of material free from pathogens and contaminants that could originate a disease and on the other there are the treatments to eliminate the inoculum in the seed or propagule. When using material free from pathogens it is important to have adequate pathogen detection methods, this is to say, sensible and specific. No production of health quality seed can be done without valid methods to detect the possibility of contamination. Some treatments can be necessary to eliminate the pathogens from the seeds, or at leafs, maintain an economically viable balance.

Health Testing of the Seed

One of the most relevant aspects about controlled health quality seed is refereed to the detection methods used on every case. ISTA (International Seed Testing Association) appeared in 1924, by was only interested in seed's gene purity and germination aspects. But detection methods were applied.

According to the criterion of each laboratory, so their results were not comparative. So, in 1957, PDC established a comparative health testing programme and standardize applied methods. The complexity for comparison of these methods made a considerable step back on the publication of working sheets (protocols). ISTA approved methods became Rules. By 1999 only 64 protocols were being compared, and only 14 of them were accepted as Rules.

As the solutions proposed by ISTA did not satisfy the needs of international seed industry, this industry, in 1994, organized The International Seed Health Initiative for Vegetables (ISHI-Veg) chartered by the vegetable seed industries in The Netherlands and France. The Initiative was soon joined by the seed companies in the United States, Israel and Japan. This group represents the production of over 75 per cent of the worlds vegetable seed supply. ISHI published many protocols, organized by crop. For horticultural crops, the ISHI-VEG, the Seed Health Testing Methods

ISHI-Veg Status

- An ISHI Validated Reference Method is a method that has been through the ISTA/ISHI comparative testing process and is being published as ISTA wn ISHI Reference Method is a method that has been through ISHI comparative testing and is under review by the ISTA/ISHI reviewers for publication as ISTA Working Sheet.
- An ISHI Accepted Method is a method that is commonly used by the seed industry for determining seed health. The method has been published in a journal, as a working sheet and/or is publicly available for comparative testing and commonly in use by the seed industry. The method is in comparative testing or aspects of the test are being tested for inclusion or deletion from the method.
- An ISHI Reviewed Method is a method that is publicly available or published, documented and prioritized by the ITG, but not subjected to a comparative test at this time. The method usually is accepted by other agencies or groups.

In 1995, during the second ISTA-PDC symposium it was decided that a Joint ISTA/ISHI Guidelines for Comparative Testing of Methods for Detection of Seed Borne Pathogens would be edited.

Through this combined effort many laboratories, organized under this alignments, the ISTA developed the Manual of Seed Health Testing Methods that has two sections:

- Validated Test Methods
- Peer Reviewed Methods.

The transition process from working protocol to ruled method is long and laborious, so working protocols can be considered as adequate methods for seed analysis. Nowadays, it is common that technological advances may become obsolete even before they are published.

The protocol to analyse *Xanthomonas campesrtis* pv.*campestris*, which causes black rot in crucifers, was published in 1982, there, it was recommended one minute to extract the bacteria from the seeds to starch culture medium. 2 years later, this time was raised to 2 hours, and other specific culture mediums. In 1987 the protocol was replaced, again in 1996 and at present time again it is under revision.

Minimum requirements to Produce Certified Seed

The Organization for Economic Cooperation and Development (OECD) establishes steps to be taken in order to produce basic and certified seed. Between these steps it is emphasized that generations from the mother material to the basic material are strictly limited in number. The number of certified seed generations, after the basic seed, varies from crop to crop and technical conditions.

In both, sexually and vegetative reproduced species, the schemes include similar steps. It generally starts with selection of a plant with genetic and health advantages. If health condition was not satisfying, certain actions should be taken in order to release it from considered pathogens. This Mother Plant is maintained under strict special security conditions in order to maintain healthiness (isolated glasshouses or chambers).

In the Certification Schemes two particularities abut seed borne pathogens should be considered, those transmitted exclusively with the seed and those that can also be transmitted with contaminated soil and plant rests. Diseases transmitted exclusively with the seed can be controlled by using clean healthy seed. This seed can be obtained by Seed Certification Programmes or by treatments that are really effective, like *in vitro* culture, thermotherapy, chemotherapy or a combined action of these, for systemic pathogens.

Tolerance levels in certified seed depend on the multiplication speed and dispersion of the inoculum of the pathogen of the considered disease. If a small amount of inoculum may be epidemic, like Anthracnose and the bean bacterial blight, if levels above zero chemical control may not be enough to stop the disease.

But pathogens whose inoculum multiplies slowly generation after generation, the disease may be evident only several generations later, so, if contamination levels are above zero may be accepted in adequate control methods are applied to the seed. This is what happens in wheat, with the Loose Smut, continuous treatments diminish inoculum levels, and make it perfectly tolerable.

New Techniques for Seed Health Analysis

Pathogen detection methods have suffered great changes since 1957. Back then it was mainly targeted to fungi, and based exclusively on incubation and identification, or the cultivation of these seeds and counting the percentage of unhealthy seedlings. These analyses were very laborious and required a high infection level in order to be detected.

Nowadays, technological advances have allowed detecting even low infection levels, evaluating 1,000 to 10,000 seeds at once. Specificity of culture mediums, usage of antibiotics and other products has allowed isolating the pathogen from the seed. Together with this, the use of reactives as mono or polyclonal antiserums and molecular markers is also available. The Lettuce Mosaic Virus (LMV), back in 1950 the symptoms were observed over many thousands of seedlings from the seed that were going to be analyzed.

Later, technical advances determined that 30,000 seedlings were needed, and even later, in 1983, an alternative to the *Chenopodium quinoa* test was discovered, the ELISA test.

Watermelon Fruit Blotch caused by bacteria is a severe problem in glasshouse winter production. It was determined that a single contaminated seed in 9,000 of them was enough too widespread the bacteria to the rest of the plants. That is the reason why screening was done over 10,000 seedlings. Nowadays PCR technique is used.

SEED SECURITY

Seed security is one of the most important issues in developing countries in particular in countries or areas frequently subject to droughts or other natural or human disasters.

CURRENT SITUATION

Many countries and international Organizations have already tried to develop seed security strategies at different levels of integration. The different strategies developed up to now are:

- Conservation by the farmers themselves of small on-farm saved seed stocks for their own needs;
- Establishment of small seed banks to be managed at village or community level;
- Establishment of national seed security stocks; and
- Development of sub-regional or regional seed security networks.

Options 1 and 2 function well in the case of a normal year, but usually fail in the case of disaster. Option 3 aimed at establishing national seed security stocks is a basic strategy in drought-prone countries, but stocks should not exceed 20-25 per cent of the annual seed requirements due to high costs to conserve such stocks.

Option 4 aimed at developing seed security networks in order to facilitate cooperation among countries in providing farmers with good quality seeds of suitable improved varieties, is certainly the best solution. Indeed, it seems obvious that achieving seed security at sub-regional or regional level is more efficient and cost-effective than at national or local level.

An indirect objective of seed security schemes is also to limit seed donations as carried out currently by many bilateral or multilateral relief programmes. Indeed, it has to be noted that in the case of emergency supply, farmers are frequently provided with seeds of poor quality (as regards germination or health) or of varieties which are not adapted to their area. In addition, such emergency supply can have adverse effects on the long-term development (i) through introduction of new pests or diseases or of new genotypes which will make crosses with local varieties or landraces and (ii) in disturbing or destroying existing commercial seed distribution networks. A good alternative would be to modify seed relief programmes in order to promote sustainable private seed sector by distributing seed vouchers instead of seeds.

Sharing Responsibilities

In the future, the role of the seed industry and seed trade to achieve seed security at regional or sub-regional level on a sustainable basis should be increased. International seed trade, seed production and seed distribution should be mainly the responsibility of the private sector (whether small, medium or large-size companies or co-operatives). Activities of seed development projects, extension services, NGOs, etc. should focus on the development of seed distribution networks based on private retailers and co-operatives, in particular in the most difficult and remote areas.

Seed security strategies should be designed at sub-regional, regional or global level, thus involving the seed industry of neighbouring countries. Pursuant to that objective, barriers to international seed trade should be removed.

SEED SECURITY: DISASTER RESPONSE AND STRATEGIC PLANNING

Since the biblical reminder in the Book of Ecclesiasticus and doubtless for antecedent millennia before, food-its shortage and its abundance-has shaped both the biological and cultural evolution of mankind.

In this modern era we are still unable to conquer the ravages of food insecurity, despite an unprecedented global capacity to produce, transport, store and process food. Daily we send news, technological know-how and money around the world in a matter of seconds. Any habitable part of the globe is less than 24 hours away and exploring the reaches of space and the depths of the oceans has become passe, yet every seventh person in the world-800 million- goes to sleep hungry at the end of each day.

If we can rid the world of smallpox, reduce polio to the status of a minor disease, almost control malaria, clean up debilitating environmental pollution, win the battle against diseases which threaten production of our crops and animals why is it that we cannot rid the world of the single most important killer-starvation? If we move genes across species barriers with impunity, transport millions of people safely across thousands of miles every year, beam the images of a world soccer cup final or the funeral of a Princess so that one person in three of four see it instantaneously, and move the natural resources of the planet from source to sink in conveyors of giant ships, why is it that we cannot produce food where it is needed, when it is needed for the people who lack a secure food supply system?

If we can build and deliver apocalyptic weapons over thousands of miles with devastating precision, launch a satellite to land on Saturn several years into the future, map the surface of the earth with unerring accuracy, build laser surgical instruments with nanometre control, why can't we predict the occurrence of events which cause large scale human disasters?

Sustainable food security rates as one the earth's most important sociological, political, economic and scientific challenges of the 21st Century.

The Nature of Food Insecurity

Food security was one of the central themes of the FAO convened World Food Summit held in Rome in 1996. FAO recognises four main categories of food insecure, hungry people:

- Poverty and chronic hunger. These are the chronically poor and comprise households with low and variable incomes, limited assets, few marketable skills, little political influence and no powerful advocates for their plight. Poverty is often associated with resource and asset poor small-scale farmers, landless unskilled labourers, livestock herders, small-scale fishermen and the otherwise unemployable.
- Life-cycle hunger. These include those who are vulnerable because of critical times in their life-cycle such as pregnant women and their unborn, the new-born, younger child-bearing and lactating women and elderly people.
- Seasonal hunger. Includes those people and households that suffer seasonal food shortages during the cycle of food growing and harvesting. They usually experience food shortage and hunger for only part of the year. Perturbations in climate can exacerbate or ameliorate seasonal hunger.
- Disaster induced acute hunger. This form of hunger results from humanitarian crisis because of natural disaster, civil and/or military strife. The problems of hunger and persistent food insecurity are compounded by frequent and excessive loss of life, loss of assets, refugees, internally displaced people and loss of societal infra-structure.

These different categories of hunger are not exclusive. Poverty, and impoverishment, pre-condition people into a state of vulnerability-vulnerable to life-cycle hunger, vulnerable to seasonal hunger and vulnerable to the impact of disaster. There is increasing evidence that a fifth group of food insecure, hungry people result from the consequences of environmental degradation and major changes in the use of natural resources. This also includes the dislocation of people from their traditional homes because of major development programmes, deforestation and large scale-exploitation of natural resources.

Complex Humanitarian Emergencies: Disaster in Chaos

The number of people affected by disaster and in need of emergency assistance rose sharply during the past decade-from 44 million people in the mid-1980s to more than 175 million by 1993. This is a 16 per cent increase per year.

A critical humanitarian disaster is a crisis in which there is a gap between the need for emergency assistance and the ability of the local government to respond.

Schell (1995) goes analytically further. He considers that a disaster is a combination of 'an extreme event and the particular dangers associated with it. A disaster occurs when individuals or entire population groups are exposed to a threat (the extreme event) in such a way that their lives are directly endangered or that so much damage is done to economic and social structures that survival becomes impossible.'

Disasters are usually thought of in terms of natural events such as drought, flood, cyclone, earthquakes, volcanic eruptions and fire. But disasters resulting from deliberate man-made crises have become more frequent and more widespread in the latter half of this century. Frequently, man-made disasters coincide with cataclysmic natural events and can often transform fragile, but manageable, environmental situations into a humanitarian disaster.

The occurrence, and frequently the severity, of these *complex emergencies* combine internal conflict with large-scale displacements of people, mass famine, and failed or failing economic, political and social institutions. In 1995 at least 26 countries were experiencing complex humanitarian emergencies or were recently emerging from long-term civil strife. An estimated 37 million people were considered to be in need of, or dependent on, international aid to avoid starvation, malnutrition and death.

A measure of the impact of complex emergencies is the number of refugees, asylum seekers and internally displaced people generated by conflict. The numbers have skyrocketed from a relatively stable 2.5 to 3 million people affected per year from 1960-1975 to an average of 38.8 million people affected per year for the period 1990 to 1995.

Notably, and tragically, more than 50 per cent of these 38.8 million people affected by complex emergencies are in sub-Saharan Africa. A microcosm of these complex emergencies is the coincident impact of civil strife and drought on the ten countries of the Greater Horn of Africa (GHA) from 1979 to 1994.

The incidence of civil strife and complex emergencies has increased noticeably since 1988. Perhaps this is in response to the collapse of the Cold War because the Cold War protagonists no longer have to 'buy and maintain ideological influence', thus allowing political and racial enmity to surface within countries and the region. In addition, vast supplies of armaments have come onto the markets at 'bargain basement prices' since the end of the Cold War.

The impact of these complex emergencies on people in the GHA. In 1994, almost 10 million people were dispossessed either as refugees or internal displacement. Most of these people live in a perpetual state of hunger and food insecurity. The magnitude of the problem in the GHA gave rise to President Clinton's 1994 Initiative on the Horn of Africa, *Breaking the Cycle of Despair: Building a Foundation for Food Security and Crisis Prevention in the Greater Horn of Africa* (US Department of State, 1994).

The increasing incidence of complex emergencies is also reflected in a greater proportion of official development assistance going to emergency relief for humanitarian aid relative to development investment in health care, agriculture, water, education and family planning. Since 1990 humanitarian aid has increased from approximately $US 1 billion to $US 3.5 billion in 1994.

Seed Security and Food Security-Preparing for Disaster

The most serious and compelling disasters are those resulting from natural, periodic occurrence of drought, civil strife and military intervention. These are the most serious types of disaster because they are usually prolonged and dispossess people from the meagre assets they have to produce their own food.

Avoiding the consequences of natural disasters involves development-type interventions such as water conserving measures in drought-prone areas and the construction of dams and other flood mitigation measures. Secure reserves of food at the local, national and regional levels is a proven way (since biblical times, and doubtless before) to preclude a serious natural disaster turning into a human calamity. In the case of complex humanitarian disasters, avoidance and preparedness is more problematic. The FAO echoes a consensus that the solutions lie in national, regional and sometimes global politics and the capacity to diffuse tensions before they erupt into crisis. While most man-made crises result from a small number of malignantly selfish and pathologically irresponsible powerful individuals, the consequences compromise the food security and lives of a great number of people, most of whom are innocent victims.

Given that disasters of one form or another are inevitable and with varying levels of predictability as to location, time, duration and course of the disaster, the regional and global community must develop a preparedness to respond to the emergency and assist the restoration of food security as expediently and sustainably as possible.

One of the 20 Priority Activities of the 1992 UN Conference on the Environment and Development (UNCED) is 'to assist farmers in disaster situations to restore agricultural systems'. The objective is to deliver seed of adapted varieties as needed to help re-establish indigenous agricultural systems in areas affected by disaster. In turn this can play a major role in restoring local food security.

SAFETY PRECAUTIONS OF SEED

Any artificial heating involves a fire hazard and this is particularly true in handling cones, since dust, resin and dry cone scales are all highly inflammable. Stringent fire precautions including a ban on smoking should be enforced, fireproof, non-wooden construction materials should be used and arrangements made for frequent removal of inflammable dust and debris by vacuum equipment or other means.

Other precautions are necessary when drying certain dry fruits and seeds of certain species. Dust masks need to be worn when handling species such as Platanus spp. which release fine hairs which might be breathed into the lungs.

SEPARATION

When fruits and cones open after drying, some seeds fall out easily as a result of manual stirring, rotation in rotating drum kilns or, in certain vertical progressive kilns, from the shaking of the cones as they fall from one tray to another. But many seeds are left inside, especially in those drying techniques where the cones remain static. They must be removed as soon as possible after drying is complete.

In some species a thorough manual shaking is sufficient to extract the remaining seed. The capsules of eucalypts need to be vigorously shaken, particularly if not fully mature, because abscission of the seeds from the placenta may be only partially complete. Failure to shake slightly immature capsules properly can result in only the chaff being released. The fertile seeds are usually attached to the placenta near the bottom of the loculus so that, after dispersal of the chaff, immature capsules may, on superficial examination, appear to be empty.

Cones may be shaken in coarse sieves to release the seeds, but more vigorous treatments are needed for some species. Those most widely used are tumbling for conifers and threshing for hardwoods.

TUMBLING

A tumbler is a rectangular or round container or drum mounted horizontally on its long axis. As it turns, cones tumble about; interior baffles often accentuate the jarring and tumbling action. Seeds fall from open cones through the high-strength wire mesh which forms the sides of the tumbler and into a hopper or trays or onto a moving belt. The tumbler can be operated by hand or mechanically driven, depending on the scale of the operation. Some drums may be closed at both ends and emptied and refilled at the end of each operation cycle (Morandini 1962). In more modern designs a continuous operation can be achieved with an inclined cylinder open at both ends. The cones are fed in at one end and during rotation roll slowly to the other end where they are discharged. Variable speeds and tilt are set for each species. The speed determines the rolling and pitching effect on the cones, while the tilt determines the length of time the cones remain in the tumbler (Turnbull 1975 c). Small types of tumbler are easily transportable. Fisher and Widmoyer describe a small tumbler of ½ bushel (18 litre) cone capacity made from a modified domestic washing machine.

In Zimbabwe a hand-operated drum tumbler 2.43 m long is used, fed from a chute from the floor above. By building the tumbling room on a slope, the cones can be easily transported into the upper floor (Seward 1980). The drum holds one bag of cones and it takes one minute to tumble one charge and to recharge the drum. The seeds drop through the 18 mm mesh of the drum into a collecting tray below and the empty cones are discharged into a trolley.

It is important that tumbling should be carried out as soon as possible after drying, because open cones exposed to cold wet air can reclose in a short time. If tumbling cannot follow immediately after drying, the opened cones should be stored in warm, dry conditions in the interval between these operations. The time required for tumbling depends on the species and the condition of the particular lot of cones being handled. Seed in some species such as Larix decidua and Picea abies may be held tightly in the cone and long periods of tumbling are sometimes required to extract them. Special machines, such as a large potato-peeling machine or a tumbler incorporating saw blades are effective on difficult species like these.

Alternatively, the cones may be rewetted and then redried to promote fuller opening of the cone scales. Haverbeke found that, after the first tumbling in Pinus sylvestris, soaking the cones in trays of water at 30°C for about half an hour until the cones softened and started to close, followed by thorough air-drying until the cone scales opened again, gave good results.

The yield of seeds from the second tumbling averaged 36 per cent of that from the first. There was a marked difference between provenances, from 18 per cent for a Scottish to 84 per cent for a Spanish origin. The additional yield from the second tumbling fully justifies the extra cost for rare and valuable seeds such as those obtained from controlled pollination.

Mechanical damage can easily be inflicted on seeds if excessive tumbling speeds are used or if the tumbler is filled with too many cones. Speed of rotation and time of treatment should be adapted to the cone and seed characteristics of the species being handled. It is better to leave some seed in the cones than to spend money on extracting seeds most of which will be severely damaged in the process.

THRESHING

Extraction of seed from dry fruits of many hardwood species is accomplished by threshing. Seed extraction in species such as Cercis, Catalpa, Robinia and Liriodendron is easily accomplished by spreading the fruits on a platform, sometimes on a straw mat or other suitable material, and beating them with a flail or slender pole. For large quantities mechanical threshers used in agriculture can be adjusted for tree fruits by altering the distance between the crushers. In chile pods of Prosopis tamarugo are ground in a stone mill set at 4 mm and the seeds are then separated by sieving and floating the milled product.

A modified cereal huller has been used as a thresher for Prosopis pods in the USA. The machine, described in Ffolliot and Thames, will thresh 1 bushel (36 litres) of pods in 1 ½ hours; approximately 160 hours would be required to thresh the same quantity by hand. In the Philippines fruits which do not readily release their seeds are put in a sack and beaten. Separation is done by use of screens. For every seed size one screen is used with a mesh larger than the seeds, to separate them from fruit fragments and other large impurities, and another with a smaller mesh which retains the seeds and allows the fine impurities to pass through.

In Sabah seeds of Acacia mangium are separated from the pods, after drying, by rotating them for 10–15 minutes in a cement mixer together with blocks of hard timber 10 × 10 × 15 cm. A similar use is made of a cement mixer for pods of Albizzia falcatariabut, because separation of the seeds is easier in this species, it is not necessary to include the timber blocks.

Several types of mechanical thresher suitable for Acacia pods are described by Doran *et al.*; they include a hand type model of resilient tapered thresher, a rotating drum, a flailing thresher and a peg drum thresher. Many acacias give off a very irritating dust during threshing and protective equipment should be worn by operators.

More robust methods, such as pounding the fruits with a wooden pestle or putting them through hammermills, must sometimes be applied. For some species special equipment has been developed such as the dehuller of Juglans. Hammermills consist of a hooded inlet or hopper, a central chamber containing a series of hammers which rotate about a central shaft, and removable outlet screens of different mesh. The outlet screen must have holes large enough to let seeds pass through without damage. Fruits are fed into the mill continuously during separation. Care must be taken to operate hammermills at comparatively low speeds, 250 – 800 revolutions per minute, to avoid injury to the seeds. The Dybvig separator works well on dry as well as on fleshy fruits.

METHODS OF EXTRACTION

Seed extraction of some species is difficult, even after standard treatments of drying and tumbling or threshing. In the Philippines indehiscent hard pods of leguminous species such as Delonix regia, Pithecellobium saman, Cassia fistula, C. javanica and Parkia javanica have to be opened with a machete or knife and the seeds picked out individually. The pods of P. saman are sweet and relished by termites; if they are piled in a dark place, after some time only the clean seeds are left behind.

Fig. The Resilient Tapered Thresher, Hand Model Made by Alf. Hannaford and Co. Ltd., Woodville, S. Australia, and used for Dry Zone Acacias.

Fig. CSIRO 15-cm Flailing Thresher, (A) Feeding Material into Thresher (B) General View Showing Material Threshed and Ready for Cleaning (C) View of Essential Parts.

Fig. Cement Mixer used for Dewinging.

Fig. *Liriodendron Tulipifera* Before and After Dewinging. Upgrading is Easier After Dewining.

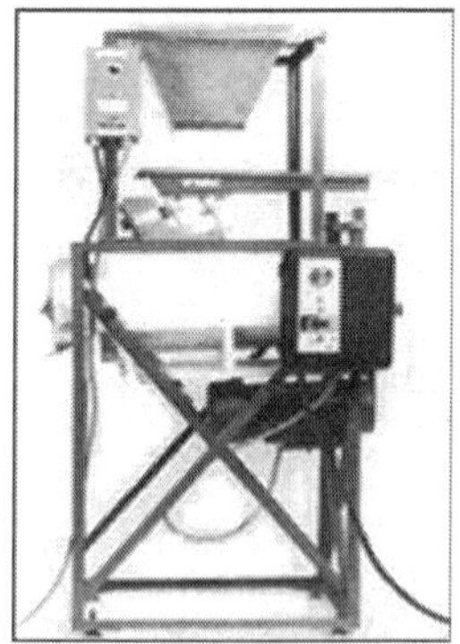

Fig. Missoula Dewinger for Small Seedlots.

Goor and Barney recommend that Cedrus seed be stored in the cones, since extracted seeds quickly lose their viability. When removed from storage, the cones must be soaked in water to facilitate extraction. After treatment the cones can be easily broken apart by hand and the seeds extracted for immediate sowing.

Serotinous cones of species such as Pinus brutia, P. halepensis, P. contorta and P. radiata may need special treatment to induce opening. Dipping them in boiling water for 10 to 120 seconds (up to 10 minutes for some especially refractory seed lots), followed by very high temperatures in the drying kiln (75° – 80° C), has been effective in some cases. High temperature is needed to melt the resin which forms a strong bonding adhesive between overlapping cone scales. Immature green cones may also need special treatment. It was found that green cones of Pinus merkusii from Zambales (Philippines), after soaking for 48 hours, followed by 80 hours' drying at a starting temperature of 30° C and a final temperature of 50° C, released only 7 per cent of the contained seeds; if the cycle of alternate soaking and drying was repeated 5 or 6 times, 79 per cent of seeds were released. The total period of 4 – 5 weeks would be uneconomic in operational forestry. The recommended alternative is to collect only ripe brown cones which released 91 per cent of their seeds in a single cycle.

OPERATIONS AFTER EXTRACTION

When seeds have been extracted from their fruits, several operations are needed before they are fit to go into storage. Sound seeds must be separated from empty and non-viable seeds and from inert fragments of fruits; winged seeds of some but not all species need to be dewinged; if seeds are to be stored, their moisture content must be tested and, if necessary, raised or lowered to the percentage most suitable for storage. If uniformity in growth of nursery stock is considered desirable, seeds may also be graded by size.

Inert material takes up space both in storage and transport and may cause uneven stocking in the nursery seed beds. It also carries a greater risk of introducing pests or diseases than do the seeds themselves; for example spores of needle cast are carried on needle fragments rather than seeds. Cleaning to a high standard of purity is easy in some species, but more difficult in others. Cleaning of seeds to a purity higher than a given percentage is undesirable in some species; beyond that, an increasing amount of good seed is separated out with the impurities. Besides, the added effort of extra cleaning is time-consuming and expensive. Morandini recommended that seeds ofLarix should not be cleaned to a purity higher than 65 per cent, since it has been shown that further cleaning causes a drastic loss of good seed.

This is because the great thickness of the seedcoat in comparison with total seed size makes empty seeds nearly as heavy as full seeds. In many species of eucalypt, especially in the subgenera Monocalyptus and Idiogenes, it is difficult to separate the fertile seeds from the chaff which is released from the capsules at the same time. Viable seeds are frequently very similar in size, shape and colour to the chaff particles and the proportion by weight of chaff to viable seeds is usually in the range of from 5: 1 to 30: 1. Commercial seed lots of eucalypts are therefore cleaned of leaves, twigs and other large fragments but the remaining "seed" consists of a mixture of seed and chaff; this practice is accepted by both buyer and seller. Provided that seed lots are accompanied by test results giving the number of viable seeds per unit weight of seed plus chaff, the seed user will not be too concerned that they contain a certain amount of impurities. Cleaning seeds for operational use should therefore be applied with discretion. It may be necessary when special techniques are used *e.g.* if the seeds are to be pelletted or if precision sowing is to be employed. For research purposes also, when basic knowledge on germination or other seed characteristics is sought, repeated cleaning to a very high standard is necessary.

DEWINGING

Winged seeds or winged fruits are a feature of many forest trees and almost all conifer seeds have a wing which may vary from long and hard to very short and soft. In order to make seed processing and nursery sowing easier, the wing is usually removed whenever it is larger than the seed (or fruit).

Wings are small or impractical to remove from seeds in several coniferous genera *e.g.* Thuja, Chamaecyparis, Cupressus; in a few genera they cannot be removed without impairing seed viability *e.g.* Libocedrus. Many winged hardwood fruits *e.g.* Casuarina, Betula,Ulmus are stored and sown intact, but the larger wings of *e.g.* Swietenia fruits may be broken off.

For small quantities of seed dewinging may be done by hand, either by rubbing seeds between the hands or against a screen or roughened surface, or by handrubbing in a cloth bag or by rolling them between two cloth sheets or in a cloth bag between a rubber surface below and a roller above. For large quantities mechanical dewinging is in common use.

Dewinging machines range from those which are hand operated to large semiautomatic equipment which gives a continuous output. Corn mixers and cement mixers are frequently used. Mechanical dewinging, if carelessly done, may cause damage to seeds by crushing, cracking or abrasion.

Most mechanical dewingers are rotating devices in which the seed is pressed by brushes or pads against the walls of a cylinder, or rotating knobs or pads force the seeds to pass through narrow apertures while retaining the wings. If the clearance between the knobs or brushes is too small the seed may be damaged. A dewinger specially designed for small seed lots of 5 kg or less and for easy cleaning between seed lots has been described by Lowman and Casavan. It consists essentially of a rubber-lined cylinder with a rotating central shaft to which are attached pure gum rubber flaps. There is a variable inclination setting so that seeds and wings will pass out of the dewinger by gravity.

Nartov *et al.* describe a combined dewinging and seed cleaning machine in use in the USSR. It is transportable and weighs 50 – 70 kg. A seed hopper and feed auger pass seeds into the dewinging unit which has brush-type beater vanes. A fan then removes wings and lighter material, while the heavier material falls onto a series of inclined screens with various mesh sizes. The clean seeds fall into different containers, graded according to seed size.

Mechanical injury can be avoided in some cases by moist dewinging. Wang describes a safe method of dewinging coniferous seeds in Canada. The seeds are moistened with water and left to soak for 20–30 minutes before they are stirred with a soft brush or sponge in a rotating cement mixer to remove the wings. A similar principle is used by Isaacs (1972) in dewinging pine seeds. A large tank with slow-moving pipes gently agitates the seeds, which have been moistened at a rate of about 2 litres of water to 45 kg of seed. The wing absorbs moisture and is shed from the seed. In Honduras moist dewinging of Pinus caribaea and P. oocarpa is done through the use of a small cement mixer or rotating drum. The capacity of the drum should be at least double the amount of seed to be dewinged and the speed of rotation about 1 revolution every 2–3 seconds (Robbins 1983a,b). The winged seeds are tumbled dry in the drum for 15 minutes; water is then sprinkled slowly over the seeds as evenly as possible while the drum continues to rotate, at a rate of about 1 litre to every 50 litres of seed.

Tumbling is continued for 45 minutes after adding the water, and then the mixture is emptied out on to a gauze-bottomed tray, after which the seeds are separated from the wings. Moist dewinging is also in common use in Sweden. In moist dewinging the seeds absorb water and have subsequently to be dried to an acceptable moisture content.

SEED MULTIPLICATION AND STORAGE

SEED MULTIPLICATION

- A continually up-dated database of multiplication technologies for each specific crop is essential and must be readily and easily accessible, particularly for asexually propagated crops where new methods of multiplication are developing rapidly.
- During and immediately following disaster there may not be the physical or human resource capacity or the opportunity to carry out seed multiplication locally. In such cases the regional programme should have developed a database of alternative multiplication sites throughout the region.
- In disasters such as drought, and where out-of-season seed production is required, a catalogue of available irrigation sites should be maintained.
- While government agencies will play an important role, the private sector also should be involved in multiplying seed specifically for seed security either by way of contracts or through some trade-off for the right to commercialise public varieties of crops.
- Also a number of NGO aid and emergency relief organisations have become involved in seed multiplication programmes to assist in seed security.
- Phytosanitary control of disease, pests and weeds in seed multiplication plots should be increased. During and after disaster fields may not be attended as assiduously as previously and crop health can deteriorate rapidly.

- If at all possible, seed quality standards should be maintained during production under disaster stress conditions.
- Planting seed acquired from outside the disaster affected region is often of poor quality and ill-adapted. Many argue that seed should be obtained as close to the disaster area as possible or grown under closely monitored contracts.
- Special attention must be given to asexually propagated species such as potato, sweet potato, cassava, other root crops and plantains for several reasons. First, multiplication near the area of need is highly preferred because of the large volume of planting material required. Second, propagules of asexually propagated crops are more prone to disease than conventional grain seed. Third, new technology to reproduce asexually propagated crops speeds the rate of multiplication and should be used if possible.
- Using local farmers to produce good quality seed can have advantages beyond supporting local seed supplies. Small area or volume production contracts can generate needed local cash flow.

Seed Storage

- Regional seed security programmes must both develop strategically located seed stores and encourage shorter term local storage on farm.
- Seed suppliers- public, parastatal or private sector- throughout the region must be obligated to maintain strategic reserves of the important crop varieties.
- Standards for seed storage are well developed and must be applied in disaster prone regions. Relevant information about seed storage can be found in Douglas (1980), Hanson (1994), FAO (1993), FAO/IPGRI (1994), WANA Seed Network (1994).
- New technologies which improve seed storage with minimal inputs are constantly under development, such as low moisture seed storage based on the use of silica gel, and should be incorporated into seed security network protocols.

STORAGE CAPACITY

The weight of seeds to be kept in store can be estimated in the manner indicated in Chapter 3 and will depend on the annual planting area, the maximum number of years' seed supply to be stored at any one time because of seeding periodicity, and the number of seeds per kg, for each species. Weight of seed in kg can be converted to net volume in litres (or in g to cm^3) by a factor related to average specific gravity. An average factor of 2.0 is appropriate for many forest species and corresponds to an apparent specific gravity of 0.5 (true specific gravity would be slightly higher because of the air-spaces between the seeds).

For conversion from net volume to gross storage space, allowing for shelving, ventilation, air spaces within and between containers, access and fittings within the cold room, a factor of about X8 is commonly used; this is with fixed shelving. Use of mobile shelving may double the quantity of seed which may be stored in a given space; in this case a factor of about X4 is appropriate. Thus 500 kg of seed of S.G. 0.5 would need a gross storage space of 500 × 2 × 8 = 8000 litres or 8 m^3 if fixed shelving were used and 4 m^3 with mobile shelving. Where relatively few seed lots and large quantities of each lot are being stored, it is possible to use standard sizes of container, each filled to the brim, and for shelving space to be adapted to fit container size exactly. Under these conditions considerable savings in storage space can be effected. Thus in the Danish seed store at Humlebaek, which uses fixed shelving, a factor of only 3.12 has been calculated.

DESIGN AND EQUIPMENT

The design and machinery for refrigerated storage is a matter for refrigeration engineers. Some guidance as to the features which should be included in any quotation for installation may be obtained from the excerpts from the IBPGR (1976) report which appears in Appendix 2 and the example of the facilities installed by the Regional Genetic Resources Project at Turrialba which appears in Appendix 3. It should be noted that both these documents refer to long-term storage of agricultural seeds for purposes of genetic conservation.

It is essential that designs and equipment be adapted to local conditions and local resources. The best installation in the world is of little use if it cannot be maintained, so it is essential to investigate the local provision for servicing and spares before committing oneself to any particular item. The reliability of mains electricity services and the need for a voltage controller and standby generator are of primary importance. Ready availability of a spare compressor may also be necessary.

The correct siting of a seed store may reduce the need for much expensive equipment. For example a tropical country with variable climate and topography might solve many problems by moving its store from a hot humid coastal site to the dry rain-shadow side of a mountain at 2000 m.

In such a case a well-ventilated room might provide perfectly suitable conditions for several years' storage for relatively "easy" species such as pines and eucalypts and could be supplemented by one or more deep-freeze chests for small quantities of more "difficult" species requiring sub-freezing temperature.

METHOD OF SEED STORAGE ATMOSPHERE

The Most obvious method of reducing the rate of aerobic respiration is to exclude oxygen from the atmosphere surrounding the seeds. This can be done by replacing oxygen by other gases such as CO_2 or nitrogen, or by using a partial or complete vacuum.

In an example with lettuce cited by Roberts, seeds were stored in sealed containers at 6 per cent MC and 18 °C. After 3 years, seed stored in an atmosphere of pure oxygen had 8 per cent viability, those in air 57 per cent, those in nitrogen or argon or CO_2 78 per cent and those in a vacuum 77 per cent. The value of excluding oxygen during storage of dry orthodox seeds has also been demonstrated in *Pinus radiata*. Best results were obtained with a storage atmosphere of nitrogen, followed by CO_2, while vacuum and air both gave poorer results.

At the highest temperature used, 35 °C, at which deterioration in viability was most rapid, the loss in final germination after 50 weeks' storage in sealed containers at 8 per cent MC was 8 per cent in nitrogen, 14 per cent in CO_2, 21 per cent in vacuum and 29 per cent in air.

The same ranking was obtained by comparing the speed of germination and the vigour of germinated seedlings (measured as dry weight 49 days after sowing). Although increases in seed longevity of this magnitude have been achieved experimentally, some of the methods are expensive to apply and the effects on seed life are less dramatic than the effects of differences in temperature and humidity. Exclusion of oxygen will prevent aerobic, but not anaerobic, respiration, whereas reduced MC and temperature will decrease the level of both.

While systematic predictions have been made of seed longevity under a range of temperature and MC for several agricultural crops, similar quantitative predictions of the effect of oxygen levels on longevity are lacking. One Simple Method which is recommended is to fill sealed containers as nearly full as possible. If there is only a small amount of air inside the container as compared with the volume occupied by the seeds, oxygen will be consumed and CO_2 produced. The resulting high CO_2/O_2 ratio is probably favourable for seed longevity in orthodox seeds.

Whereas the complete exclusion of oxygen from the storage atmosphere appears beneficial to most dry orthodox seeds, there is evidence that some oxygen is necessary for recalcitrant seeds. Seeds of *Araucaria hunsteinii*, which had an initial germination of 56 per cent, had all died within a month if stored in pure nitrogen, in two months if stored in 1 per cent oxygen, and in three months if stored in 5 per cent oxygen, while germination was still 18 per cent after four months' storage in 10 per cent oxygen (Tompsett 1983, 1984).

Storage in a polythene bag of 25 micron thickness, which was ventilated (21 per cent oxygen) periodically when opened for extraction of samples, gave results very similar to those in 10 per cent oxygen. King and Roberts record a general consensus that adequate ventilation (*i.e.* adequate oxygen), is necessary for the successful storage of recalcitrant seeds at relatively high MC, as well as for storage of imbibed seeds of orthodox species.

SEED MOISTURE CONTENT

The Relationships of seed moisture content on a wet weight or fresh weight basis to seed MC on a dry weight basis, and of the equilibrium moisture content of seeds to the relative humidity of the surrounding atmosphere, are important in seed processing. They are equally important in seed storage. In the first case manipulation of RH can effectively change MC of seeds to the optimum for storage, in the second case MC can be maintained at or near that optimum by maintaining a suitable RH in the atmosphere around and between the seeds.

EFFECT OF MC

In orthodox seeds, moisture content is probably the most important single factor in determining seed longevity. Reduction in MC causes a reduction in respiration and thus slows down ageing of the seed and prolongs viability. Harrington, cited by barner, has related MC to various processes within and around the seed as Follows:

Seed Moisture Content % (wet weight)	
Above 45 – 60 %	Germination begins
Above 18 – 20 %	The seed may heat (due to a rapid rate of respiration and energy release)
Above 12 – 14 %	Fungus growth can occur
Below 8 – 9 %	Insect activity much reduced
4 – 8 %	Sealed storage is safe.

Prevention of fungal activity is more easily achieved by controlling MC than by controlling temperature. If MC and RH are high enough, fungal activity is possible between –8 °C and +80° C and it is easier to keep MC below 12–14 per cent (or RH to the equilibrium of around 65 per cent) than to maintain sub-zero temperatures.

Within the range of 4 to 14 per cent MC, Harrington has suggested a rule of thumb applicable to many agricultural species—the life of the seed is doubled for every 1 per cent decrease in MC, Schönborn found a relationship of a similar order when measuring the respiration rate, expressed in terms of CO_2 production, of *Picea abies*. At 20 °C and 20 per cent MC the seed gave off 80 ml. CO_2 per hour per kg of seed; at 20 °C and 5 per cent MC, the rate of CO_2 production was reduced to 0.11 ml/hr/kg, a reduction of nearly a thousand times for a difference of 15 percentage points in MC.

An MC of 4–8 per cent is considered safe for most orthodox species; 5 per cent ± 1 per cent is recommended for long-term storage for genetic conservation. Oily seeds will usually tolerate drying to a somewhat lower moisture content (calculated on the basis of total fresh weight) than will non-oily seeds. Drying below 4 per cent can lead to damage or more rapid loss of viability in some species, although certain species can be dried to a considerably lower MC. *Betula papyrifera* was successfully stored at 0.6 per cent MC without injury (Joseph 1929 cited in Holmes and Buszewicz 1958), while Schönborn (1965) succeeded in drying small samples of Picea abies, Pinus sylvestris, *Pseudotsuga menziesii* and *Larix decidua* down to 0 per cent MC without any observable drop in germination after 6 months, compared with germination at the normally applied 6–8 per cent MC. Drying in this case was done, not by exposing the seeds to high temperatures but by leading a current of dry air through the seeds at 20 °C.

The Same treatment killed *Pinus strobus* and *Abies alba,* and earlier attempts by Barton to store seeds of several species of *Pinus* and *Picea* at 0 per cent MC also failed. Below about 2 per cent MC desiccation injury becomes a strong possibility in many species. Drying to very low MC is also more costly than drying to the usual 4–8 per cent and is likely to be used only in exceptional cases. Some orthodox forest trees store best at appreciably higher MC. 8–10 per cent is recommended for *Fagus sylvatica*. For seeds of *Abies* spp., an MC of 12–13 per cent is recommended for storage of one to three years, but for longer periods this should be reduced to 7–9 per cent. Species which benefit from storage at higher than average MC also need particular care in the timing and speed of drying.

Fluctuation in the moisture content of seed in storage due to open storage without humidity control or to frequent opening and resealing of sealed containers results in deterioration in the germinability of the seeds. In fact a steady MC slightly above the optimum is usually less harmful than one which fluctuates between the optimum and a higher moisture content.

Some cases have been reported in which the usual trend of decreasing seed longevity associated with increasing MC is reversed at or near the moisture content of fully imbibed seeds. If the species in question needs exposure to light in order to germinate, it is possible to store fully imbibed but ungerminated seeds for some time in the dark. *Fraxinus americana* was Stored at 22 °C at varying MCs with the following results:

MC%	**Germination (%) after Storage Periods Indicated**			
	1 month	2 months	3 months	4 months
6.0	98	92	96	94
9.5	94	88	76	4
18.6	81	22	0	0
Fully imbibed (in dark)	96	95	98	96

It Has Been postulated that imbibed seeds can repair damage to cell membranes, enzymes and DNA in the cell nucleus caused by free radicals in a way which is not possible for seeds at lower MC. Prolonged imbibed storage may, however, be difficult in practice, because of the need to maintain constant high moisture for imbibition and adequate oxygen without allowing the seeds to germinate or encouraging the multiplication of fungi and bacteria. Moisture content is also important in recalcitrant seeds, but in this case the critical MC is the minimum to which it is allowable to dry the seeds rather than the maximum content for prolonged storage. For many of the large temperate hardwood seeds, moisture contents in the range of 25–79 per cent are appropriate.

Storage should be carried out at close to the minimum safe MC, since the higher the MC the higher the respiration rate and the more rapid the loss of viability. Higher respiration rates release higher amounts of energy and there is a risk of overheating and death of the seed, unless great care is taken to provide adequate aeration. High MC also increases fungal activity and the spread of rot. Wang quotes results from two seed lots of Acer saccharinum; germinability of one lot stored at 58 per cent MC and 1–2 °C dropped from 94 per cent to 12 per cent after 6 months' storage, while that of the other, stored at 45 per cent MC at the same temperature, was still 78 per cent after 16 months. Loss of viability in this species may be sudden. Tylkowski found that seeds stored in sealed bottles at 50–52 per cent MC and –1° to –3°C had over 90 per cent germination after 18 months but this dropped to nearly zero after 24 months.

Less Research has been done on tropical recalcitrant species, but there is some evidence *e.g.* in *Triplochiton,* that viability may be significantly prolonged if the minimum MC for the species can be determined and if particular care is taken over the period of drying down to that MC. Trials of *Shorea platyclados* in Malaysia indicated that a gradual reduction of MC to 20–27 per cent, followed by sealing in charcoal, sawdust or vermiculite at 15°–22 °C allowed storage for at least one month, compared with the week or so of natural viability.

On the basis of experiments carried out on *Shorea parvifolia* and *Dipterocarpus humeratus,* Maury-Lechon *et al.* recommended reduction of MC to between one quarter and one half of the initial MC in freshly collected fruits. Although tropical recalcitrant seeds cannot yet be stored for more than short periods, there is a growing body of useful research on the problem.

PURPOSE OF SEED STORAGE

The ability of seed to tolerate moisture loss allows the seed to maintain the viability in dry state. Storage starts in the mother plant itself when it attains physiological maturity. After harvesting the seeds are either stored in ware houses or in transit or in retail shops. During the old age days, the farmers were used farm saved seeds, in little quantity, but introduction of high yielding varieties and hybrids and modernization of agriculture necessitated the development of storage techniques to preserve the seeds.

Seed storage is the maintenance of high seed germination and vigour form harvest until planting. Is important to get adequate plant stands in addition to healthy and vigourous plants. Every seed operation has or should have a purpose.

The purpose of seed storage is to maintain the seed in good physical and physiological condition from the time they are harvested until the time they are planted. Seeds have to be stored, of course, because there is usually a period of time between harvest and planting. During this period, the seed have to be kept somewhere. While the time interval between harvest and planting is the basic reason for storing seed, there are other considerations, especially in the case of extended storage of seed.

Seed suppliers are not always able to market all the seed they produce during the following planting season. In many cases, the unsold seed are "carried over" in storage for marketing during the second planting season after harvest. Problems arise in connection with carryover storage of seed because some kinds, varieties, and lots of seed do not carryover very well.

Seeds are also deliberately stored for extended periods so as to eliminate the need to produce the seed every season. Foundation seed units and others have found this to be an economical, efficient procedure for seed of varieties for which there is limited demand. Some kinds of seed are stored for extended periods to improve the percentage and rapidity of germination by providing enough time for a "natural" release from dormancy.

Regardless of the specific reasons for storage of seed, the purpose remains the same maintenance of a satisfactory capacity for germination and emergence. The facilities and procedures used in storage, therefore, have to be directed towards the accomplishment of this purpose.

In the broadest sense the storage period for seed begins with attainment of physiological maturity and ends with resumption of active growth of the embryonic axis, *i.e.,* germination. Seeds are considered to be physiologically and morphologically mature when they reach maximum dry weight. At this stage dry-down or dehydration of the seed is well underway. Dry-down continues after physiological maturity until moisture content of the seed and fruit decreases to a level which permits effective and efficient harvest and threshing.

This stage can be termed as harvest maturity. There usually is an interval of time between physiological maturity and harvestable maturity, and this interval represents the first segment of the storage period. Any delays in harvesting the seed after they reach harvest maturity prolongs the first segment of the storage period–often to the detriment of seed quality.

The second segment of the storage period extends from harvest to the beginning of conditioning. Seed in the combine, grain wagon, and bulk storage or drying bins are in storage and their quality is affected by the same factors that affect the quality of seed during the packaged seed segment of the storage period. The third segment of the storage period begins with the onset of conditioning and ends with packaging. The fourth segment of the storage period is the packaged seed phase which has already been mentioned. The packaged seed segment is followed by storage during distribution and marketing, and finally by storage on the farm before and during planting.

The control that a seedsman has over the various segments of the storage period for seed varies from a high degree of control from harvest to distribution, to much less control during the postmaturation-preharvest, distribution-marketing, and on-farm segments.

Despite variable degrees of control over the various segments of the storage period, the seedsman's plans for storage must take into consideration all the segments. The things that can be done must be done if the quality of the seed is to be maintained.

METHODS

The conditions which prolong viability during storage have been well defined for seeds which are tolerant of desiccation. Storage conditions have been recommended by the IBPGR Advisory Committee on Seed Storage. For base collections, seeds of between 3-7 per cent moisture content should be stored in sealed containers. Sub-zero temperatures are acceptable, but –18 °C or less is preferred. For active collections sealed storage of seeds dried to 7 per cent moisture content or less is recommended at temperatures of less than 15 °C. Unsealed storage is not encouraged. In particular, it is not recommended in tropical areas.

CHECK THE NUMBER OF SEEDS IN THE ACCESSION

- Weigh the seeds of each accession.
- Convert the weight of seeds to seed number, using the thousand seed weight of each accession for an accurate conversion. An approximate conversion can be done on a species basis by using the approximate weights given in Appendix 2 of Cromarty, Ellis and Roberts (1982).
- For accessions containing mixtures of genotypes, the sample size should be at least 4000 seeds. For genetically uniform accessions, the sample size should be at least 3000 seeds.
- If the sample contains more than the required number of seeds, proceed to storage.
- If the sample contains less than the required number of seeds, either proceed directly to regeneration or store the seeds temporarily in the genebank and regenerate at the earliest opportunity.

Notes and Examples

An approximate inter-conversion of seed number and weight can be done easily using the thousand seed weight.

Example for Sorghum (1000 seed weight = 17 g) 100 g contains:

$$\frac{1000}{17} \times 100 = 5750 \text{ seeds}$$

These are minimum sample sizes for the start of storage and if both space and seeds are available, more seeds should be held.

Regenerate as soon as possible. Seeds processed and stored under good conditions will not loose viability before regeneration.

Equipment

Coarse balance

DETERMINE WHERE THE SEEDS SHOULD BE LOCATED

- Check the inventory data file of the genebank to find the next available space where a container can be located.
- When seeds from the same regeneration cycle of the same accession are stored in several containers, keep all the containers of the accession together.
- Make a list of where each accession will be placed.

Notes and Examples

The storage arrangements will vary among genebanks. The most important point is to know exactly where to locate each accession within the store.

Equipment

- Racks
- Trays, boxes or drawers
- Coldroom or freezer

PLACE IN THE SEED STORE

- Place the container into the seed store in the listed location.

ENTER THE DATA INTO THE DATA BASE

- Fill in the data on the location and date of storage of each accession and each container into the data file.
- Record the date of the next monitoring test for germination in the data file. This date will be determined by the curator after considering the viability and moisture content of the seeds, storage conditions and the IBPGR recommendations.

Notes and Examples

A code can be used to locate an accession within the store. Each unit can be identified by a number or letter in the store. The code can indicate the number or the letter of the freezer, store, rack, basket or drawer, etc.

Example

A010201 could be used to indicate the location as:

Coldroom:	A
Rack:	01
Shelf:	02
Box:	01

Colour codes can also be a quick and easy way to locate accessions. A colour can be used for each rack, shelf or species. This both speeds up the work in the coldroom and makes it easy to spot errors. Owing to the very cold temperatures, the faster that one can locate accessions in the cold room the better.

SEED SHIPMENT

The benefits of exemplary seed collection, processing and storage methods may be largely lost if care is not taken over shipment from seed store to nursery. It is seed viability at the time of sowing, rather than at the time of despatch from the seed store, which determines the number of healthy plants produced from a particular seed lot. It is therefore essential to provide shipment methods which will ensure the minimum loss of viability in the interval between storage and sowing. The selection of appropriate packing material will depend on the characteristics of the species, the quantity to be shipped, the length of time in transit, the mode of transport and the temperature and moisture conditions to which the shipment will be exposed (Baldwin 1955).

High and fluctuating temperatures and adverse humidity are the chief causes of viability losses during shipment (Stein *et al.* 1974). These factors are identical with those that cause deterioration in freshly collected fruits between the collecting site and the processing depot. However, seeds between storage and sowing should start with advantage of having had optimum conditions of temperature and moisture content during the storage period. In fact, maintenance of storage conditions during transit would be ideal, but is often not possible.

Provided that the initial moisture content of the seeds is correct, it can be easily maintained during transit by the use of sealed containers. In some cases the seeds can be despatched in the same containers in which they were stored. In others it may be advisable to transfer them from a large container in storage to a smaller container for despatch. Individual nurseries may require only a small quantity of a given seed lot. In addition, small and light packages are often less subject to mechanical damage in transit than large, heavy ones. Magini (1962) recommends separate packages of 1 – 20 kg but not larger. A variety of moisture-proof or moisture-resistant material is available, as described earlier in this chapter under storage containers. Polyethylene of 4–

8 mil (100–200 microns) has the advantage of restricting moisture passage while allowing exchange of oxygen and CO_2.

Sealed containers are highly suitable for orthodox species, of which the seeds must be kept dry during transit. The addition of a desiccant such as silica gel may be a useful additional insurance if there is any risk that the seeds may absorb moisture while being transferred from storage container to shipment container. Seeds of recalcitrant species, on the other hand, are best left unsealed, since the effect of some loss of moisture is less harmful than that of the overheating which can occur as a result of rapid respiration in sealed bags at ambient temperatures. They should be well-mixed with pulverized sphagnum moss, ground peat, coconut fibre or sawdust, that has been moistened and squeezed dry. A mixture of equal weights of dry packing and water will give adequate moisture content to these materials. In the case of international transit, however, an inert non-organic substance such as moist vermiculite is likely to be more acceptable to quarantine authorities.

Sealed moisture-proof containers should always be used for long journeys, *e.g.* form one country to another, of orthodox species of short longevity, provided that the initial MC is correct. But if orthodox seeds are being forwarded soon after collection and without having been dried to the appropriate MC for storage, it is preferable to ship in bags permeable to air rather than to seal with excessively high MC. A number of species with resistant seedcoats or pericarps, such as Tectona and many leguminous species, are able to withstand prolonged periods in ambient conditions; cotton or paper bags or hessian sacks are perfectly suitable for these species.

Large, moist seeds can be sealed individually with paraffin wax or latex. In the method described by Baldwin (1955), paraffin wax is heated to 71°–77°C and seeds or nuts dipped for a few seconds in a screen-type container, which should be shaken vigorously during the immersion. The waxed seeds should be packed in soft material so that the wax is not scraped off during transit. At the time of sowing the wax must be partly scraped off to permit the entry of water.

Protection against high or rapidly fluctuating temperatures is more difficult, but care should be taken to avoid placing the seeds close to local hotspots such as radiators and hot pipes. For very sensitive seeds, temperature effects can be mitigated by the use of insulating material in the packaging.

Sub-zero temperatures do not usually affect dry seeds but may cause damage to recalcitrant seeds which must be kept moist. Premature germination is another risk which affects moist seeds. During storage, germination can be restricted by the use of low temperatures just above freezing, but the higher temperatures encountered during transit may induce germination in a substantial number of seeds. Seeds which are prone to germinate when held in moist packing may be treated with an inhibitor such as maleic hydrazide.

No matter what type of seeds is being despatched, it is necessary to take precautions against mechanical damage to seeds and against losses due to damage to the containers in transit.

Double wrapping is often advisable, for example a sealed polythene bag should be placed inside a stout canvas bag. Stout drum cartons with sealed polythene or aluminium-foil containers inside provide an especially effective combination for seeds which need to be kept dry. If the inner bag is labelled, this is also an insurance against accidental defacement of the outer label. Clear labelling is essential and the consignee should be advised of despatch by means of an appropriate seed consignment note or seed issue form.

Stein et al. (1974) have provided a useful check list of helpful practices in seed shipment, reproduced hereunder:

- Double wrap the seed. Enclose the seed container in a sturdy, preferably rigid, outer container.
- Small or moderate size containers generally withstand shipment better than large containers.
- Fill containers completely to minimize air content and jostling of seeds during shipment.
- All packages should bear a good identifying label on the innermost covering and another one within the container.
- For long distances, shipment of sensitive seeds by air is desirable.
- Seed packages should permit ready opening and reclosing if destined for export to a country requiring fumigation. In addition a copy of the phytosanitary certificate should be readily available to quarantine authorities *e.g.* by sealing it in an envelope which is firmly attached to the outside of the package.

Seed storage facilities at nursery sites or district forest stations are inferior to those at the central seed store. Shipments should therefore be timed so that seeds can be sown with the minimum delay after receipt.

A BASIC SCENARIO FOR A SEED SECURITY STRATEGY

In March 1996, a Consultative Meeting of representatives of NARS, IARCs, NGOs and inter-governmental agencies was held in Entebbe, Uganda, to develop an action plan which would provide a continuing technological response to disaster by re-establishing food security through rapid replenishment of adapted varieties of the major food crops in countries of the Greater Horn of Africa (GHA).

The Consultative Meeting concluded that four basic activities were necessary to implement a regional seed security programme.

DEVELOPING NATIONAL SEED POLICIES

Unless there is a formally structured National Seed Policy (NSP), preferably by enactment of a Seed Act, it is difficult for a country to put in place measures which would enhance the level of seed security. Without a frame of reference provided by a NSP it would be even more difficult for a country to negotiate and come to agreement on a regional seed security programme.

A NSP is equally necessary for seed security to enable response to disaster and for the orderly and healthy economic development of a seed industry during non-disaster periods.

A regional seed security programme should develop a detailed set of principles to assist individual countries to establish or improve their respective national seed policies. The significant components of a national seed policy must include:

- Plant improvement and variety development, including access to advanced germplasm and genetic resources from IARCs and plant breeding programmes in developed countries.
- A routine system to facilitate variety evaluation, registration and release which adopts policies and protocols which are consistent with worldwide practice.
- The development and enactment of legislation for Plant Variety Rights which is consistent with the model law developed by UPOV, the International Union for the Protection of New Varieties of Plants.
- Some form of seed certification is necessary with emphasis on quality, purity and phytosanitary cleanliness. Effective seed certification schemes require adequate training and mechanisms to accredit and monitor personnel and facilities.
- The establishment and enforcement of reasonable quarantine measures to protect against the uncontrolled movement of pests, diseases and weeds across country borders.
- Promotion of the production, storage and marketing of high quality seed domestically. A strong private seed sector usually leads to effective self-regulation of seed quality.
- A National Seed Policy is virtually essential for the orderly establishment and maintenance of strategic seed reserves.
- An NSP must recognise and promote the informal on-farm seed production system in its overall scheme of seed industry development and seed security. There is no doubt that in Rwanda the informal seed sector kept seed distribution channels open long after the formal seed sector had ceased to operate. On-farm seed production systems, on-farm selection, the maintenance of old varieties, land races and mixtures, local and community seed storage and the operation of local seed marketing and distribution channels are vital aspects of agriculture in developing countries.
- A formally constituted National Seed Advisory Committee or

Agency appointed at Ministerial level is usually necessary to oversee the implementation of an NSP and to update the seed policy as is necessary.

Regional Harmonization of Seed Standards and Regulations

Regional policies that facilitate the movement and exchange of seeds across country borders is absolutely essential for effective regional seed security.

This was certainly the experience of the member countries of SADC where collaboration among the countries of the region averted the occurrence of a major human tragedy resulting from the 1991/92 drought.

The issues that must be resolved at the regional level to enable countries to come to agreement on how to harmonize seed standards, regulations and the orderly movement of seed include:

- Certification of varietal purity. The OECD has established procedures to certify varietal purity of seed that enters international trade.
- Standards and procedures for seed testing. The International Seed Testing Association (ISTA) has established procedures which promote uniformity of testing and ISTA also certifies seed testing facilities. In some countries compliance with the ISTA code for seed certification is mandatory, in others it is voluntary.
- Plant protection and quarantine regulations. The FAO has established a set of guidelines that are widely adopted. Special provision has to be made for asexually propagated crops such as potato, sweet potato and cassava because of the increased risk of disease transmission from one area to another using vegetatively propagated material.
- Plant variety protection. Plant Variety Rights (PVR) have been adopted by many developed countries consistent with the model law and guidelines developed by UPOV (1995). Recent World Trade Organisation discussions deal with the proposed requirement for countries trading in agricultural commodities to have some form of intellectual property protection of plant varieties- Trade Related Intellectual Property Protection (TRIPP).
- Major debate continues about PVR (Crucible Group 1994). Issues are added seed cost, reduced number of varieties, restriction on flow of germplasm to plant breeders, restrictions on farmer's privileges to save and exchange seed, compensation of developing country farmers for germplasm already contributed to variety development, restrictions on trade and marketing, and relationship of PVR and the Convention on Biological Diversity.
- Seed marketing. The sale of seed is frequently dependent on a form

of seed certification and in some countries seed growers and seed sellers must be licensed or registered. Internationally acceptable guidelines have also been developed to provide some degree of uniformity in contractual arrangements for seed sales.

- Regional strategic seed reserves. The creation and maintenance of regional strategic seed reserves would be the epitome of regional cooperation to respond to disaster. Regional reserves could be an extension of national strategic reserves but would require explicit agreement and networking among member states.
- Regional associations and networking. Networking is essential as the first step for member countries to develop a regional seed policy. Increasingly the countries in a region, related primarily by geographical proximity but also by economic imperative, are forming regional associations. In part this is a response to the evolution of country affiliations away from ideological association toward common economic opportunities, and regional security and development.

SEED TRADE AND SEED SECURITY IN DEVELOPING COUNTRIES

SEED SUPPLY IN DEVELOPING COUNTRIES

Seed supply in developing countries relies on two distinct sources:

- the seed industry, be it public or private, and
- on-farm seed, which is the main source, particularly in the most remote areas.

The Seed Industry

It is difficult to collect data on the commercial seed market in developing countries. However, it is possible to say that the commercial market represents less than 10 per cent of the seed consumption, in particular for the staple food crops.

This is due to several factors, but the most important one is certainly the weakness of the seed distribution networks. Due to the lack of involvement of the private sector, the number of seed suppliers is not sufficient. In some countries, seed production and distribution is a public sector monopoly.

On-farm Seed

On-farm seed production is the main source of seed in developing countries, in particular for small-scale farmers (SSF) in the most remote areas.

For some food crops such as sorghum, millet, cowpea, cassava and sweet potato, on-farm seed or planting material represents more than 90 per cent of seeds used by the farmers. It must be noted that this percentage is an average, since figures vary a lot according to climatic conditions: after droughts, farmers have no or little on-farm seed available and, therefore, require access to commercial seed or seed provided through relief operations. On-farm seed must be considered as a major component of seed supply and seed security in developing countries, but it is not a stable one. Consequently, it is difficult and very risky to develop a sustainable seed security strategy based only on such a component. The two following diagrams show (i) the evolution of seed sources depending on the years (normal year *vs.* year with calamity disrupting national production) and (ii) that the 'informal seed sector' is a weak component to ensure sustainable seed security:

STRATEGIES FOR SEED SECURITY IN DEVELOPING COUNTRIES

Seeds provide the greatest good at minimal cost. Simply, seeds stand between survival and starvation.

Arguably, seeds have the greatest socio-economic benefit to human welfare of any known biological device. They are, first and foremost, the source of all food, at least of plant and animal origin.

Seeds concentrate really useful technology into the most transportable, the most storable, the most nutritional, the most tradeable, the cheapest and the most functional format that could possibly be imagined. Seed delivery programmes are vitally important to maintaining and improving agricultural productivity not only in the developing world but also in the industrialised world. Though not a comprehensive list, a wider appreciation of the importance of seeds and their crucial role in agricultural and therefore human development, can be gained by referring to the books and monographs published by Simmonds (1979), Douglas (1980), Kelly (1989), Heiser (1990), Cromwell et al (1993), Srivastava and Jaffee (1993).

The increasing incidence of complex humanitarian emergencies, sometimes coincidently with natural disaster, has escalated the cost of emergency food aid. Thus, it is little wonder that establishing seed security is seen as a fundamental to maintain and restore food security. A number of recent initiatives and analyses are addressing the issue of the relative role and importance of seed security as a first defence against food insecurity. These initiatives and analyses have tended to focus on sub-Saharan Africa because of the greater incidence and severity of complex humanitarian emergencies. They include:

- A joint Commonwealth Secretariat and SADC workshop in 1994 to recommend action by countries of the SADC region which would strengthen seed programmes and harmonise regulations, laws and the institutional framework in order to promote the production of good quality seed and to facilitate regional trade (Commonwealth Secretariat and SADC 1994).
- A review of the literature to examine the role of seed regulatory frameworks and their adequacy or otherwise to help current

attempts to improve the performance of the seed sectors of developing countries.

- A USAID promoted study to provide information that will be useful to agencies planning and implementing seeds for disaster mitigation and recovery programmes for farmers who have suffered from natural or complex disasters in the Greater Horn of Africa.
- A review by the Overseas Development Institute of the latest developments in knowledge and techniques in emergency seed distribution and related long-term seed capacity building activities, and to stimulate discussion as to what constitutes 'good practice' in this field.
- A USAID sponsored feasibility analysis including a consultative workshop and development of an action plan that will provide a continuing technological response to disaster by re-establishing food security through rapid replenishment of adapted varieties of major food crops in the Greater Horn region of Africa.
- A World Vision initiative- Food Security for Africa Programme- has grown into a bold initiative, 'The Year of the Seed', involving crop improvement and seed supply projects in 15 new countries in Africa during 1997.
- A World Bank initiative, promoted by ASARECA, to gather information and evaluate the current constraints to the movement of seed across borders of ASARECA countries with the objective of 'Standardising Seed Policies and Regulations in Eastern and Central African Countries'.

From these various initiatives and studies a consensus is developing about the strategy and types of activities that have to be implemented in order to establish and maintain seed security in developing countries.

The same basic ideas also apply to restoring seed security following natural disaster or complex humanitarian emergencies. The dozen or so consensus ideas distilled from the above studies and analyses, ranked more or less in relative importance, are as follows:

- The Development of Regional Databases and Information Exchange Networks to include such information as varieties and their distribution, geographical and agroecological adaptation, aliases, reserve quantities and quality, descriptions and relevant data that characterise and distinguish crop varieties; technical and other information about seed storage, multiplication, quality and phytosanitary certification and regulations; lists of seed growers, cleaners, merchants, importers and exporters; and, other available data and information.

  Of note is the recent development of Crop Environment Domain (CED) Maps. These crop-specific agroecological maps, which are based on Geographic Information Systems (GIS), integrate adapted crop variety and land race characteristics with agroclimatic regimes. CED maps have considerable value to assist in seed security and to restore seeds of the best varieties following disaster.

- Establish Regional Coordination Networks to foster a strong and continuing partnership among all the players concerned with maintaining seed security and/or restoring it after disaster. This includes national agricultural research systems (NARS) and other relevant national government agencies, regional organisations, non-government organisations (NGOs), inter-governmental organisations, (IGOs), private volunteer organisations (PVOs), IARCs, the private and public seed sector, farmers, communities and their respective associations, and international donors. One of the reasons for the success of the Seeds of Hope initiative in helping to restore seed security to Rwanda was the rapidity of the collective and coordinated response of the neighbouring NARS, eight IARCs, several NGOs, UN agencies and donors.
  Similarly, the successful response to the 1991/92 drought in southern Africa depended on rapid and effective collaboration among a similar set of organisations operating in and among the ten drought-afflicted countries.
- Development of Plans to Produce and Access Seed, both to maintain seed security and to restore seed security in the event of disaster. Aspects that need to be taken into consideration include local varieties and land races; best bet adapted varieties from the region and beyond; areas of similar agroclimatic conditions or with irrigation outside the affected region for emergency seed increase; quality and quantity of seed required; mechanisms to ensure timely delivery of seed; contract protocols to acquire high quality seed; to minimise profiteering and to help ensure timely delivery of the right seed to the right place.
- The Establishment of National Seed Policies are essential for the operation of a secure seed system. A National Seed Policy would include breeding and plant improvement objectives and strategies; variety testing and registration of varieties through a Variety Approval and Release system; referee seed testing and certification, including affiliation with the International Seed Testing Association; implementing reasonable and pragmatic quarantine systems; licensing and regulation of the seed industry and its members; legislation for the protection of new varieties of plants that is consistent with the UPOV (International Union for the Protection of New Varieties of Plants) convention; and provide official representation to regional and international bodies.

- Harmonising and Standardising Seed Regulations and Policies across a Common Region to increase access to the best regional technology, to increase regional trade of seed and promote and speed up the exchange and movement of seed between countries. There are technical issues to be considered including phytosanitary regulations, quarantine, accreditation of laboratories, field operations and facilities.
- Promoting the Establishment and Long-Term Maintenance of National and Regional Strategic Seed Reserves. This has already been implemented with more or less success in several countries such as Ethiopia, Morocco, Cyprus, Zimbabwe, South Africa, Botswana and Bangladesh. Governments must encourage the establishment of national strategic seed reserves, possibly through an imaginative approach involving private sector seed companies.
- Regional Training, Accreditation and/or Licensing of Seed Quality Personnel and Agencies to ensure that minimum standards are observed and maintained within countries and across the region. Professional competence is important and minimum requirements of training need to be negotiated by the countries of a region. Similarly, it is necessary to establish a mutually recognised mechanism to license seed certification and phytosanitary inspectors, seed testing laboratories and facilities.
- Seed Transport, Delivery and Distribution Systems to ensure that the best quality seed of the right type gets to the more vulnerable farmers and communities at the right time. This is most important to maintain a level of seed security during disaster or to restore seed security after disaster. Frequently, the national agricultural programme and country infrastructure is disrupted and R&D oriented IARCs do not have the capacity to deliver and distribute the right seeds to needy people at the right time.

  Rather, NGOs are better equipped to implement delivery and distribution and are often more aware of specific needs. Experience and post-intervention impact assessment in southern Africa and Rwanda revealed that the local seed distribution and marketing system seemed to continue to function during and immediately after the disaster.
- Farmer and Local Community Ownership and Empowerment is a critically important component to re-establish seed security, local food security and even political stability. As the Rwanda experience clearly identified, farmers and their communities are the lynch-pin in re-establishing and maintaining seed distribution systems. Similarly they can be pivotal in re-establishing local seed production, cleaning and storing of short term seed supplies. A catalytic injection

of seed supplies, equipment, technology and possibly some funds may be required. With local community ownership and empowerment, such support may well prime cash flow and asset creation within the community.

- Rebuilding the R&D Capacity to Sustain Future Seed Security is an important follow-up once the basis of seed security has been established or re-established. If there is no seed security system in place, building a R&D capacity is a major developmental programme. If the R&D capacity, including personnel and facilities, is more or less degraded as a result of disaster, early intervention can resurrect that capacity relatively quickly.

  In Rwanda, the SOH programme recognised the need to provide seed and breeding material, equipment, limited funds to pay technical staff and minimally refurbish facilities, to retrain staff or to train newly recruited staff, and to provide relevant knowledge and guidance. This had a significant impact on assisting the Institut des Science Agronomique du Rwanda (ISAR) to rebuild its own seed security infrastructure relatively quickly.
- Evaluating the Impact of an Intervention to Restore Seed Security should be a component of any effort to restore seed and food security following a disaster. The problems encountered, the solutions found and the lessons learnt from the course of the disaster and the manner in which the intervention to restore seed security unfolded should be documented and analysed. Such an analysis provides case examples, guidance, denial or confirmation of preconceptions and good general experience to aid future interventions.
- Exiting from an Emergency Seed Programme following successful restoration of seed security can be problematic. In the very tragic situation of Rwanda, food security and internal R&D capacity was essentially lost. The Seeds of Hope programme became a lifeline not only to help restore seed security but also to help rebuild R&D capacity. The very nature of the success of the Rwanda SOH intervention led to a form of mutual dependence among the SOH partners and ISAR. SOH could have become a surrogate for conventional long-term development assistance. SOH extended the original deadline by six months to complete the intervention during which time Rwanda began the process of restoring conventional development programmes.

It is important that an emergency intervention to restore seed security establish clear and measurable milestones at the outset and that these be carefully adhered to. It is also important that the programme establish a time when the intervention should cease.

3

Process of Seed Germination

Seed germination is a process by which a seed embryo develops into a seedling. It involves the reactivation of the metabolic pathways that lead to growth and the emergence of the radicle or seed root and plumule or shoot. The emergence of the seedling above the soil surface is the next phase of the plants growth and is called seedling establi-shment.

Three fundamental conditions must exist before germination can occur:

1. The embryo must be alive, called seed viability.
2. Any dormancy requirements that prevent germination must be overcome.
3. The proper environmental conditions must exist for germination.

Seed viability is the ability of the embryo to germinate and is affected by a number of different conditions. Some plants do not produce seeds that have functional complete embryos or the seed may have no embryo at all, often called empty seeds. Predators and pathogens can damage or kill the seed while it is still in the fruit or after it is dispersed. Environmental conditions like flooding or heat can kill the seed before or during germination. The age of the seed affects its health and germination ability: since the seed has a living embryo, over time cells die and cannot be replaced. Some seeds can live for a long time before germination, while others can only survive for a short period after dispersal before they die.

Seed vigour is a measure of the quality of seed, and involves the viability of the seed, the germination percentage, germination rate and the strength of the seedlings produced. The germination percentage is simply the proportion of seeds that germinate from all seeds subject to the right conditions for growth. The germination rate is the length of time it takes for the seeds to germinate. Germination percentages and rates are affected by seed viability, dormancy and environmental effects that impact on the seed and seedling. In agriculture and horticulture quality seeds have high viability, measured by germination percentage plus the rate of germination.

This is given as a per cent of germination over a certain amount of time, 90 per cent germination in 20 days, for example. 'Dormancy' is covered above; many plants produce seeds with varying degrees of dormancy, and different seeds from the same fruit can have different degrees of dormancy. It's possible to have seeds with no dormancy if they are dispersed right away and do not dry (if the seeds dry they go into physiological dormancy). There is great variation amongst plants and a dormant seed is still a viable seed even though the germination rate might be very low. Environmental conditions effecting seed germination include; water, oxygen, temperature and light. Three distinct phases of seed germination occur: water imbibition; lag phase; and radicle emergence. In order for the seed coat to split, the embryo must imbibe (soak up water), which causes it to swell, splitting the seed coat. However, the nature of the seed coat determines how rapidly water can penetrate and subsequently initiate germination.

The rate of imbibition is dependent on the permeability of the seed coat, amount of water in the environment and the area of contact the seed has to the source of water. For some seeds, imbibing too much water too quickly can kill the seed. For some seeds, once water is imbibed the germination process cannot be stopped, and drying then becomes fatal. Other seeds can imbibe and lose water a few times without causing ill effects, but drying can cause secondary dormancy.

Inducing Germination

A number of different strategies are used by gardeners and horticulturists to break seed dormancy. Scarification which allows water and gases to penetrate into the seed, include methods that physically break the hard seed coats or soften them by chemicals.

Means of scarification include soaking in hot water or poking holes in the seed with a pin or rubbing them on sandpaper or cracking with a press or hammer. Soaking the seeds in solvents or acids is also effective for many seeds. Sometimes fruits are harvested while the seeds are still immature and the seed coat is not fully developed and sown right away before the seed coat become impermeable.

Under natural conditions seed coats are worn down by rodents chewing on the seed, the seeds rubbing against rocks (seeds are moved by the wind or water currents), by undergoing freezing and thawing of surface water, or passing through an animal's digestive tract.

In the latter case, the seed coat protects the seed from digestion, while often weakening the seed coat such that the embryo is ready to sprout when it gets deposited (along with a bit of fertilizer) far from the parent plant. Microorganisms are often effective in breaking down hard seed coats and are sometimes used by people as a treatment, the seeds are stored in a moist warm sandy medium for several months under non–sterile conditions. Stratification also called moist–chilling is a method to break down physiological dormancy and involves the addition of moisture to the seeds so they imbibe water and then the seeds are subject to a period of moist chilling to after–ripen the embryo.

Sowing outside in late summer and fall and allowing to overwinter outside under cool conditions is an effective way to stratify seeds, some seeds respond more favourably to periods of oscillating temperatures which are part of the natural environment. Leaching or the soaking in water removes chemical inhibitors in some seeds that prevent germination. Rain and melting snow naturally accomplish this task. For seeds planted in gardens, running water is best—if soaked in a container, 12 to 24 hours of soaking is sufficient. Soaking longer, especially in stagnant water that is not changed, can result in oxygen starvation and seed death. Seeds with hard seed coats can be soaked in hot water to break open the impermeable cell layers that prevent water intake.

Other methods used to assist in the germination of seeds that have dormancy include prechilling, predrying, daily alternation of temperature, light exposure, potassium nitrate, the use of plant growth regulators like gibberellins, cytokinins, ethylene, thiourea, sodium hypochlorite plus others. Some seeds germinate best after a fire, for some seeds fire cracks hard seed coats while in other seeds chemical dormancy is broken in reaction to the presence of smoke, liquid smoke is often used by gardeners to assist in the germination of these species.

Origin and Evolution

The origin of seed plants is a problem that still remains unsolved. However, more and more data tends to place this origin in the middle Devonian.

The description in 2004 of the proto–seed Runcaria heinzelinii in the Givetian of Belgium is an indication of that ancient origin of seed–plants. As with modern ferns, most land plants before this time reproduced by sending spores into the air, that would land and become whole new plants. The first "true" seeds are described from the upper Devonian, which is probably the theater of their true first evolutionary radiation. The seed plants progressively became one of the major elements of nearly all ecosystems.

DEVELOPMENT OF GERMINATION

When the pod of the bean is developing, the embryo in the seed is being fed by the parent and visibly grows until ripeness is attained. The young plant then assumes a dormant or resting state within the seed without showing any signs of life. Under certain conditions, however, the plantlet begins to wake up, and soon escapes from its protective coat to lead a separate and independent life. This awakening from a resting condition to a state of active growth is called germination and is dependent upon an adequate supply of

- Water
- Heat
- Air or oxygen.

It is also essential, of course, that the plantlet in the seed must be alive. The exact nature of the dormant state of seeds is not understood, but in old seeds and those which are gathered in an immature condition or badly stored the embryo is often weakened or actually dead; in the latter case no germination is possible.

The exact length of time which seeds may be kept before death of the embryo takes place has never been satisfactorily determined; it varies with the species of the seed, its ripeness and composition, and also with the method of storage. In the case of most farm and garden seeds kept in the ordinary way, few of them are found capable of growth after ten years, and a large number die in two or three years.

For present purposes it will suffice merely to mention that age is a determining factor in the germination of seeds. Water is necessary is well known, as beans may be kept indefinitely in a sack or drawer at various temperatures and with access to air without germination taking place.

When placed in moist ground, or between damp blotting-paper, they absorb water very readily. This is most easily observed when beans are soaked for twelve hours in a dish containing water. The water is transmitted through all parts of the coat, but much more quickly and easily through the micropyle and the line of softer material which runs the whole length of the centre of the hilum.

It is rapidly brought into contact with the part of the embryo which grows first, namely, the radicle. The soft spongy thicker part of the inside of the testa lying beneath the hilum stores up a considerable amount of water for the benefit of the developing plant, and the whole of the embryo and the seed-coat absorb water and become softer and larger in consequence; it is only after this swelling has happened that a bean begins to show any signs of germination.

The need of an adequate temperature for germination is a matter of common knowledge among those accustomed to sow seeds. If soaked beans are placed in the ground in midwinter they show little or no signs of waking from their dormant condition, yet when placed under a glass on damp blotting-paper indoors, the radicle makes its exit from the seed in a few days.

Seeds differ in the temperature which is necessary to induce them to germinate, the embryos in some commence to extend their radicles and push their way through the seed-coat even if just kept above freezing point; others require a temperature of 9° or 10°C to start growth.

If attempts are made to grow beans at 45°C it will be found to be too hot, and they make little or no progress. Between this high temperature at which growth appears impossible, and the freezing point where the development of the embryo of the bean is also suspended, there is a temperature at which the embryo makes the most rapid progress, and emerges from the seed-coat in the shortest possible time; this most favourable temperature is about 28°C, both above it and below it the germination of the bean is retarded. The supply of fresh air is also an essential condition for growth of the young plant from the bean seed, but the evidence for its need is not so manifest or so generally recognised as the necessity for moisture and warmth.

It will be found, however, when beans are placed in a flask or bottle containing carbon dioxide or hydrogen gas they refuse to germinate, even when they are supplied with a proper amount of water. The peculiar extension or growth of the parts of the interior of the bean seed, and the fact that a suitable supply of water, air and heat is necessary for the manifestation of these changes, suggests to us that we are dealing with a living structure.

This becomes all the more«apparent when we observe that the oxygen of the air is absorbed, and in its place carbon dioxide is given off into the surrounding air, for this is what happens in the breathing of a living animal. Carbon dioxide is produced when beans germinate.

Place twenty soaked beans in a wide-mouthed bottle, and cork them up after showing that a match burns freely in the bottle. Leave them in a warm place for twenty-four hours, and try if a match will now burn in the bottle. The carbon dioxide gas can be poured out into a beaker containing lime water; on shaking, its presence is proved by the lime water becoming 1 milky ' owing to the precipitation of carbonate of lime. The particular use of the water, heat and air to the plant we cannot at present discuss.

Without water the embryo would have little chance of becoming free from the tough and hard seed-coat surrounding it; water softens the latter, and makes it more easily torn by the extending radicle and plumule. In the early stages of the life of the bean plant, from the commencement of germination up to the time when the first green leaves are unfolded, the development and building up of the elongating rootlet and shoot depend upon the thick cotyledons.

At first the latter are thick and fleshy, but as the radicle and plumule grow the cotyledons become softer and thinner, ultimately shrivelling considerably. The cotyledons are leaves, the interior of which is packed with food for the rest of the growing embryo, and a large amount of the water absorbed by the seed is used for the purpose of dissolving the nutrient material in them, and carrying it from them to the various parts of the root and shoot of the young plant where growth is going on the cotyledons are esential to the development of the root and shoot of the embryo by cutting them off as soon as the two latter parts have emerged from the seed-coat. Try separating one cotyledon and then two at various stages of development, and see if the axis (root and shoot) can be made to develop without them. The growth should be allowed to continue some time in order to obtain well-marked effects. Not only do the changes observed innhe embryo of a germinating bean point to the conclusion that it is a living structure and like an animal dependent on a proper supply of water, heat and air for the manifestation of its life, but the parts of a young bean plant after emerging from the seed soon give evidence of the possession of peculiarities which are associated with life. When put in the ground, the radicle, in coming out of the seed, turns straight downwards and continues to grow in this direction. This is the case no matter in what position the seed is placed. If, after germination has commenced, it is taken and replanted with the primary root pointing to the surface of the soil, the tip of the root soon begins to curve downwards again, and will maintain this course until again disturbed.

The plumule behaves in exactly the opposite manner; after emerging from the seed-coat its bent tip grows upwards and away from the root; if the seed is reversed and replanted the plumule begins to curve in such a manner that its tip is driven upwards towards the surface of the ground. That these peculiarities are somehow connected with life is clear, as dead embryos show no such behaviour. Sow soaked beans in a flower pot or box filled with ordinary garden soil placing them in various positions in it, some laid on the flat side, some with the hilum directed upwards, and others with the hilum downwards. Allow them to grow in a warm place: take them up as soon as signs of germination are noticed, and observe the direction the root and shoot have taken. The peculiar tendency for the root always to go downwards and stem upwards can be investigated by sowing beans in ordinary garden soil and afterwards reversing them.

To avoid error all should be taken up, and then placed again in the soil in various positions—some as they were, a few with their roots and stems reversed, and others laid in a horizontal position. They may be re-examined at the end of a week. When the roots have extended about half an inch take two seeds and suspend them by means of thread side by side in a bottle with their roots downwards and stem upwards. The bottle should contain a little water to keep the air damp.

When the roots have grown about two inches reverse one of the seeds so that its root points upwards tad stem downwards. Notice that the tip of the root of the reversed seed in about twelve hours begins to turn downwards, while the plumule more slowly bends in such a way as to assume the position it had before it was reversed. The bottle should be placed in a dark box or cupboard to avoid the influence of light on the plant, and fresh air should be blown into the bottle twice a day. Although seeds vary almost indefinitely in regard to size and shape they are similar to the bean in so far as they all contain a young plant packed away within the seed-coats.

In this essential feature all seeds agree with few exceptions, and it is on account of the existence of a young plant within them that they are of use in the raising of crops or plants. The manner, however, in which the embryo is arranged, and the relative size and appearance of its various parts, differ considerably in seeds; moreover, the growth during and after germination is not the same in all cases. A few of the more important and common variations in these respects must be noticed.

THE COMMON BEAN

A broad bean is one of the largest seeds met with in ordinary farm or garden practice, and as its parts are all sufficiently large to be observed without the special aid of anything more than an ordinary pocket lens, it is especially fitted for study. When a nearly ripe pod of a broad bean plant is opened, each seed within it is found attached to. The inside by means and it is through this stalk that all the nourishment passes from the parent to enable the young seed to develop.

A1 first the pod exists in a rudimentary form in the centre of a flower and its parts and contents are very small; they are nevertheless readily seen with a pocket lens. After the fading of the flower, the pod and seeds within it grow larger and larger at the expense of food supplied by the rest of the plant, and ultimately when ripe the funicles wither and dry up, and the seeds become detached from the parent which has produced them. When dry and ripe each bean seed is hard, with an uneven surface, but its internal construction cannot be clearly examined in this condition.

On soaking in water for twelve hours, however, it becomes softer, and the parts can then be easily investigated. The outside, which is a pale buff colour, is smooth, and has at one end a narrow elongated black scar called the hilum of the seed. It is known popularly as the ' eye' of the bean, and marks the place where the broad end of the funicle separated from the seed when it ripened in the pod. Quite close to one end of the hilum is a very minute hole known as the micropyh, easily seen with a lens, and through which water oozes out usually accompanied by bubbles of air when soaked beans are squeezed between the finger and thumb.

This opening communicates with the interior of the seed, and is the only one it possesses. The rest of the seed after the testa is removed, is of oval flattened shape similar to the complete bean, and is divisible into two large fleshy halves called cotyledons which, however, are not completely separate from each other, but connected at the side with a conical projecting body, one end of which is found to fit into a hollow cavity in the seed-coat exactly opposite the micropyle; the other end is bent and turned inwards between the fleshy cotyledons. The extent and shape of this small curved structure is most easily observed when one of the cotyledons is removed completely; it remains attached to the other. Soak some broad beans in water and keep them in a warm place all night. Examine them next day and make drawings of the various parts seen both before and after stripping off the testa. Observe the relative position of the parts of the embryo in reference to each other and to the seed-coat.

Examine and compare the structure of the following seeds after soaking in the same way:—Pea, scarlet runner beans, vetches, and red clover. The bean seed contains nothing more than what has already been described; the nature and relationship of its component parts only become intelligible when the seed is placed in the ground or maintained under certain conditions, and allowed to grow. When growth commences the lower end of the small curved structure elongates and breaks its way through the coat of the seed at a point very close to the micropyle, but not, as often erroneously stated, through the micropyle itself.

It soon assumes the form and is recognised as a root of a young bean plant The upper bent half, which lies between the cotyledons, also pushes its way out of the same opening in the seed-coat and develops into a stem, from the tip of which leaves are gradually unfolded.

It is thus seen that the seed of a broad bean is a packet containing a bean plant in a rudimentary condition. This plantlet is callai an embryo, and the portion of it which becomes root and stem is its primary axis.

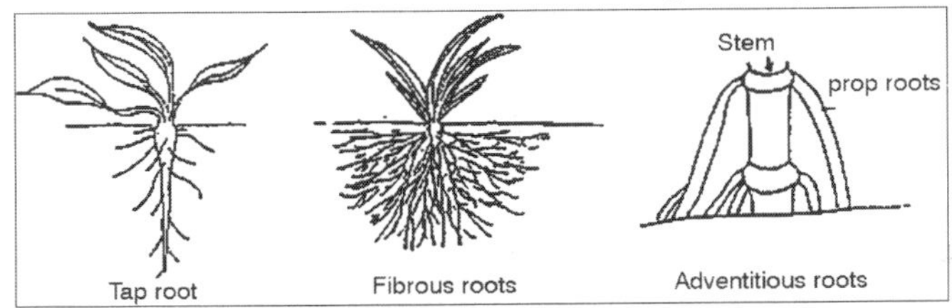

Fig. Root System

The part of the primary axis which is below the point where the cotyledons are attached consists of a very small piece of stem, the hypocotyl at the end of which is the radicle or primary root. Where the stem ends and the root begins cannot be determined in the bean seedling without the aid of the microscope and examination of the internal structure of the axis of the plant. The curved end of the primary axis above the cotyledons is the plumule of the embryo, and consists of a very short piece of stem, the tpicotyl on the top of which is a bud.

From the latter is derived the ordinary stem which comes above ground with its green leaves and flowers. In the early stages of the growth of the embryo from the seed the hypocotyl grows very little. The upper part of the stem bearing the plumule comes out of the seed bent and it maintain this curved shape for some time after emerging. By this behaviour the delicate leaves of the plumule are protected from injury during their progress upwards when a seed is sown in earth or sand. Fold up some soaked beans in two thicknesses of white flannel made damp, and place them on a plate. Cover them with another plate placed upside down, and leave them in a warm room.

Examine them twice a day, leaving them, exposed to the fresh air for a few minutes each time, and keeping the flannel damp, not wet. When they sprout notice the place where the radicle has come out of the seed-coat. Let some grow till the radicle and plumule are well out of the seed, and compare the various parts of the sprouted seeds with unsprouted ones.

SEED OF WHITE MUSTARD

The seed of white mustard (Brassica alba Boiss.) contains an embryo which like that of the bean consists of a radicle, plumule and two cotyledons; the latter, which are folded together, are relatively thinner than those of the bean and deeply notched. The radicle is bent round and lies in the fold of the cotyledons, between which is the very small, almost invisible plumule. On germination the cotyledons, instead of remaining within the seed-coat and below the ground as in the case of the broad bean, escape from the enclosing coats altogether and grow up out of the ground, enlarging at the same time, and becoming green like ordinary leaves.

They are the first ' smooth' leaves of the seedling mustard plant. After a short time the plumule grows up from between the cotyledons and forms a stem upon which are gradually unfolded the ordinary divided ' rough' leaves. Soak white mustard seeds, and examine their structure, noting especially the way in which the embryos are packet in them.

Allow some to germinate and grow for a week or more on damp flannel, and examine them in various stages of development, noting the notched cotyledons with small. The term hypogean is applied to cotyledons which remain below ground, those coming above being epigean, the relative amount of growth in the hypocotyl and epicotyl determining their position. If the hypocotyl grows vigourously during or after germination the cotyledons are forced above ground; when only the epicotyl grows the plumule is lifted up above the soil, but the cotyledons remain below where the seed is placed. In the broad bean the hypocotyl is very short, and the point where it ends and the root begins is not clearly denned. In a mustard seedling, however, the point of separation between the root and stem is somewhat swollen and readily distinguished.

All plants whose embryos, like those of the bean and mustard, possess two cotyledons, are known as Dicotyledons they form a very large, well-marked class of the flowering or seed-bearing plants. The seeds contain within their coats nothing but an embryo plant, which depends for the development of its root and shoot upon the substances stored up in some part of its body, its cotyledons chiefly. This is true even in the case of seeds like those of white mustard, in which the cotyledons of the embryo are comparatively thin. There are, however, a number of plants, such as the ash, mangel and potato, which, although belonging to the Dicotyledons, have seeds in which there are stores of food inside the seed-coat, but free from the embryo and its cotyledons. Such separate reserve-food is stored in that part of the seed known as the endosperm or 'albumen,1 and seeds in which it is present are called endospermous or albuminous seeds.

Those like the bean, pea, and vetch, mustard, and turnip, which have no separate reserve-food, are known as exendospermous or exalbuminous seeds. Take out a seed from the fruit of the ash tree in autumn; carefully cut thin shavings from the flat side of the seed, stafting about the middle of the seed and cutting towards the narrow end. Note the white embryo with its well-marked radicle, hypocotyl and two flat cotyledons lying within the semi-transparent endosperm. Some of the most commonly occurring endospermous seeds will be found to have embryos within them which are not dicotyledonous, and whose structure is in many respects. A good example is met with in the onion.

WHEAT GRAIN

A wheat grain, which may be taken as an example, is not a seed, but a kind of nut with a single seed within it. The seed grows in such a way as to completely fill up the interior of the nut, and become practically united with its inside wall. The embryo occupies only a small part of the grain, the rest being taken up by the floury endosperm of the seed. The embryo is easily seen at the base of a soaked grain on the side opposite the furrow.

When removed it has the appearance. The part of it which lies close up to the endosperm is a flattened somewhat fleshy shield-shaped structure called the scutcllum attached to the front of the scutellum is the plumule, consisting of a bud formed of an extremely short stem, upon which are sheath-like leaves enclosing each other.

The embryo generally possesses five roots, one primary and two pairs of secondary. They are all completely enclosed by a sheath which continuous wjtn the scutellum, and are consequently not visible from outside; their position, however, is marked by projecting bosses. The sheath round the roots is termed the coleorhiza, and when germination takes place it expands and bursts the coats of the grain, the roots about the same time breaking through the enclosing coleo-rhiza.

When a wheat grain is sown in the ground it remains there, but the plumule grows upwards, its first leaf, the coleoptile, appearing above the soil as a single pale tube-like structure; from a slit in the tip of the latter the first flat green blade soon appears, and is followed by a succession of single green leaves, the younger ones growing from within the older ones in regular order. With a sharp knife or razor cut through from back to front, so as to divide the grain into two longitudinal halves, and note the floury endosperm and the shape and parts of the divided embryo. Place a folded sheet of damp blotting paper on a plate, sow some soaked wheat grains on it, and cover with a tumbler. The grains will germinate. Watch their development up to the time the first green leaf appears, taking out the embryo and examining it at different stages of its growth. There is difference of opinion as to which part of the embryo is to be considered the cotyledon.

Soine authorities regard the scutellum as the cotyledon, while others give this name to the coleoptile or first sheathing leaf which comes above ground, and which has no green blade. Others, again, consider that the first sheathing leaf is an extension of the scutellum, and the two combined is therefore the cotyledon. In any case, there is only one cotyledon present, and wheat therefore belongs to the class of monocotyledonous plants. During the growth of the embryo of a wheat grain, it will be noticed that the endosperm becomes soft and decreases in quantity as the roots and plumule expand and develop; the endosperm is the food upon which the young plant depends during the early staggs of its life, the scutellum acting as a structure for changing, absorbing, and transferring this reserve-food to the growing parts which need it. Remove the embryos from well-soaked grains, and grow them without the endosperm on damp blotting-paper.

Allow ordinary uninjured grains to grow with them. Both the embryos in the grains and those removed from the grains develop, but there is a great difference in the results after a few days.

The store of reserve-food on which the young plant depends for its early development is sufficient to enable it to form a root, stem, and several leaves, as is evident when seeds are allowed to germinate upon damp flannel or blotting-paper, from which nothing but water is absorbed. No food-materials or manures are needed for this primary development, and seeds germinate and the seedlings grow for a considerable time as well in poor soil or sand as in good rich ground.

As soon as the reserve store is exhausted hunger becomes apparent, and unless the plants are then supplied with suitable nutriment from the soil and air, and are also placed under conditions favourable for growth, weakness and death are likely to occur. Among the larger seeds, such as beans and peas, where there is an abundant store of reserve-food, the young seedlings begin to manufacture food for themselves from materials absorbed from the soil and air, long before their reserve is exhausted.

In small seeds the reserve is sometimes almost consumed before the roots and leaves are sufficiently developed to carry on their work properly, in which cases a more or less temporary starvation and check to growth ensues. Especially does this happen when seeds are sown too deeply, for a large amount of food is then used in the production of a stem long enough to lift the leaves up into the air.

ONION

The seed is black, somewhat oval in outline, with one side convex, the other almost flat. Each contains within it endosperm and an embryo which lies curled up inside in the form seen. When germination commences, the curved part imbedded in the middle of the endosperm grows and forces the end of the embryo out of the seed.

From this exposed end, which is the radicle, a straight, slender, primary root develops. The part of the young seedling which extends from the root into the interior of the seed, grows very rapidly at first, at the same time assuming a sharply bent outline.

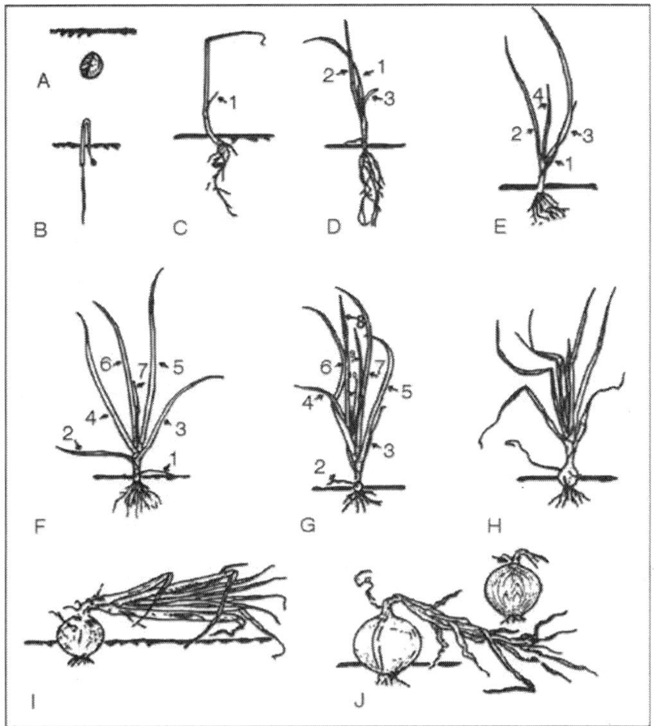

Fig. Germination of Onion

It comes above ground in the form of a close loop but on further growth the end within the seed is pulled out of the soil, and grows up in the air. The tip within the seed changes and absorbs the endosperm, and usually remains there until all the nutrient material has been transferred from it to the various centres of growth in the young plant. After the food-reserve is exhausted the tip withers and becomes free from the seed coat.

In loose soils the latter is pulled above ground before the endosperm is exhausted, and remains on the end of the tip for some time. In other cases where the soil is damper and of a stiffer nature the seed-coat remains below ground altogether. The curved part of the embryo which comes above ground is a leaf. It is the cotyledon of the embryo, and is in reality a thin hollow leaf like those of the full grown onion plant: within it is the plumule, which consists of a series of hollow conical leaves arranged one inside the other. After one leaf emerges others soon follow, the younger ones coming out in regular order through slits in the sides of those immediately older than themselves. Soak fresh onion seeds in water for a few hours. With a razor cut through some parallel to their flat sides in order to show the embryo within.

Sow others in damp blotting-paper; allow them to germinate and the seedlings to develop; make observations of them at different stages of growth. Watch the germination of seeds sown in boxes or pots containing ordinary garden soil. Plants whose embryos possess only one cotyledon are known as Monocotyledons, and form the second large class of seed-bearing plants. Few of the representatives of this class with which we are ordinarily familiar have true seeds large enough for examination.

BEGINNING OF GERMINATION

Germination takes place only in the first week after the beginning of the summer rains, and the seedlings are hidden in the litter under bushes or among the small stones which cover the ground.

The saguaro is a subtropical species and its seeds germinate only in the summertime. Germination occurs principally in July and August during the southwestern summer monsoon. Thus, successful germination and early establishment depend on a short period of coincident high levels of moisture and warmth during but a few weeks of the desert year. We are not surprised, therefore, to find that this highly successful desert species has acquired adaptive strategies that fit a set of germination requirements to an environment characterised by a persistently high evaporation which is only partially offset by scanty and often uncertain rainfall.

Seedfall is completed just prior to the normal arrival of summer rains. In exceptional years, the two events—seedfall and rainfall—actually overlap. The first summer rains contribute importantly to seed survival by dispersing seeds and by washing them into locations that offer concealment from predators. Little or no germination, however, results from these late June and early July rains. Rigid light and moisture requirements interact to inhibit germination and prevent losses that would result from germination at that time, for sporadic early monsoon rains seldom provide a sufficiently reliable supply of moisture to promote germination and support the initial establishment of seedlings. The increasing frequency of rains during the second half of July offers progressively more favourable conditions for seed germination.

While it is commonly said that "seedlings germinate," plant seedlings "sprout" and proceed to establish or die; only seeds can germinate. The highly variable spatial and temporal distribution of summer rainfall importantly affects the number of saguaro seed germinations occurring at a given location. Further, the number of germinations varies in accord with the characteristically large year-to-year variability of monsoon precipitation patterns. Our experimental and field observations, however, indicate that the most favourable conditions for saguaro seed germination do not necessarily occur during years with the highest total summer precipitation.

Rather, natural germination is associated with temporal clustering of summer rainstorms that provide continuously high moisture levels at and near the soil surface during the 2- to 3-day germination period. Most germination takes place during the period from mid-July through the first week of August. Then, with continuing availability of high surface moisture resulting from short-spaced rainstorms and supplemented by dewfall, germination proceeds to completion within a period of 2-3 days—initial establishment of the seedlings takes place at a time when temperatures and moisture are optimum for their growth.

The high temperature limitations on saguaro seed germination operate to restrict natural germination to shaded sites where periods of high moisture are sufficiently prolonged to permit germination. During wet periods such sites seldom reach temperatures high enough to inhibit germination. Inhibition of germination by low temperatures effectively restricts germination to the warm-wet periods of June, July, August, and September in the overall distribution of the saguaro.

Table. Per cent Soil Moisture in Shaded and Unshaded Saguaro-paloverde Habitat at Saguaro National Monument (east), July 1967

Date July	Soil moisture (%)			
	Shade		Open	
	0.5 inch (1.27 cm)	1 inch (2.54 cm)	0.5 inch (1.27 cm)	1 inch (2.54 cm)
18	15.23	15.34	6.01	8.12
19	10.06	9.63	3.13	6.69
20	3.40	8.87	2.36	5.45
21[a]	1.85	6.33	1.39	4.47
22	11.45	9.96	7.80	7.54
23	5.15	8.10	2.43	4.50

[a]Estimated by regression (log % on time).

Samples of the top 0.5 inch (1.27 cm) and top 1 inch (2.54 cm) of soil were collected at approximately 1300 hr beneath the crown of a mature foothill paloverde at 1 m (3.3 ft) from the trunk (shade), and from an unshaded (open) level site approximately S m (16.4 ft) outside the crown. Data graphed in Figure.

Seeds were broadcast beneath the crown of a foothill paloverde (Cercidium microphyllum) on 16 July immediately prior to receiving 0.20 inch (5 mm) precipitation the same day. No germination occurred during the period of high soil moisture following a heavy rain the next morning, followed by clear skies (17 July, 1.5 inches; 38 mm). Germination took place after two periods of light precipitation (0.04 inch; 1 mm, 0.21 inch; 5 mm) and broken overcast skies on 21 and 22 July. The first seedlings emerged on 23 July, 3 days after the start of these rains.

Amount (bar height) and period (bar width) of precipitation is shown. Saguaro seed germination is dependent upon the continued availability of moisture at the soil surface as determined by temporal distribution of rainfall, cloud cover, and other factors that aid in continuously maintaining high relative humidity and surface moisture for a period of 2-3 days during the summer rainfall season. Generally, natural germination is associated with the occurrence of two or more rains within a 2- to 3-day period. Genetically controlled variability in the time required for completion of the germination process further insures that the entire annual seed crop will not respond to conditions that are marginal for seedling establishment.

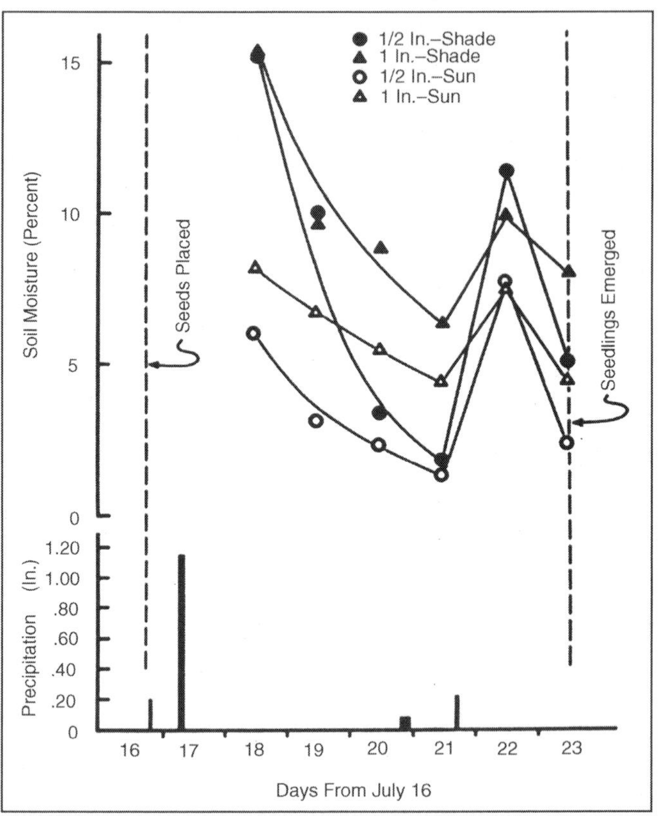

Fig. Natural Field Germination of Saguaro Seeds (on 23 July 1967) and Accompanying Precipitation (inches [Original units of Measurement; 1 inch = 25.4 mm]) and Soil Moisture at 0.5 inch (1.27 cm) and 1 inch (2.54 cm) depth.

Approximately 10 per cent of seeds tested under ambient summer light and temperatures required 4 days or more of continuous exposure to free water for completion of germination. Thus, even under conditions that permit extensive germination, a substantial portion of the germinable seeds remain viable and available for subsequent germination.

The light requirement may seem to present an apparent survival disadvantage because it seemingly dictates that germination must occur at or near the soil surface where the predation pressures are greatest and requisite moisture conditions are least apt to prevail. However, in view of the characteristic development pattern of the monsoon rains, an important adaptive strategy can be recognised. In the presence of insufficient moisture to complete germination and initial establishment of the seedling, the light requirement acts to stop the germination process without loss of seed viability. Buried seeds retain their viability, and will germinate quickly when exposed by subsequent rains, animal digging, other natural forces, or experimental exposure.

The light requirement insures germination sufficiently near the soil surface for the globular seedling to emerge and immediately receive adequate light for subsequent growth. Sensitivity to far-red light may also permit better utilisation of energy from nocturnal re-radiation thus hastening completion of the germination process. Field observations indicate that in nature germination of shallowly buried seeds accounts for a substantial portion of the annual saguaro seedling crop. This observation appears to conflict with knowledge that germinating seeds must receive exposure to light after they have imbibed water, that little or no germination (<1 per cent) will occur in total darkness. Extensive laboratory tests using a variety of environmental, mechanical, and chemical treatments have revealed no mechanism by which germination might occur naturally in the absence of light (Alcorn and Kurtz 1959; McDonough 1964). Based on successful use of acid treatments in promoting dark germination, McDonough (1964) suggested that passage through the digestive tracts of animals might promote germination in the absence of light. However, we have attained no success in the dark-germination of saguaro seeds recovered from the feces of native rodents or birds.

In fact, the natural germination of buried saguaro seeds requires no complex mechanism for explanation. In view of the brief exposure and low intensities of light required for germination as found by all investigators, the natural translucence and gravelly surface of the soils of the saguaro habitat will admit sufficient light in most instances to permit germination of seeds buried to a depth of 5-10 mm (0.2-0.4 inches), the approximate maximum depth from which the seedling can emerge. A further mechanism for satisfying the light requirement exists in the soil churning and washing action occurring during typical summer rain storms. Such action permits burial of seeds after wetting and exposure to light.

Germination of saguaro seeds requires contact with free water and exposure to light (Alcorn and Kurtz 1959; Alcorn 1961a). Under ambient summer light and temperatures in a laboratory environment, initial germination of saguaro seeds maintained in continuous contact with free water occurs approximately 48 hr after initial wetting and 50 per cent germination is reached in approximately 72 hr.

However, within the natural environment of the saguaro such prolonged periods (3 days) of continuously available water are a rare, almost non-existent occurrence during July and August when most natural germination takes place. Rather, natural germination commonly follows two or more shorter periods of water availability with interspersed drying conditions at the soil surface resulting from the occurrence of two or more distinct rainstorms within a 2- to 5-day period (Steenbergh and Lowe 1969). In a humid atmosphere, saguaro seeds hygroscopically imbibe and retain moisture required for germination.

Such pre-exposure to high relative humidity effectively speeds the germination process and reduces the required period of seed contact with free water. First germination of seeds so pre-conditioned can take place after 24 hr of wet contact and 50 per cent germination can occur within 48 hr. In a near-saturated atmosphere, maximum hygroscopic imbibition (approximately 20 per cent of the air-dry seed weight) is reached in approximately 20 hr.

Table. Hygroscopic Imbibition by Saguaro Seeds in the Laboratory at High Relative Humidities (90-100 per cent).

Time (hr)	Weight (g)	Uptake (%)
I - Ambient Light		
0	2.00	0.0
1	2.10	5.0
2	2.16	8.0
4	2.23	11.5
6	2.26	13.0
17	2.35	17.5
19	2.36	18.0
21	2.32	16.0
23	2.30	15.0
25	2.29	14.5
II - Total Darkness		
0	2.00	0.0
12	2.36	17.8
15	2.37	18.5
21	2.41	20.5
23	2.41	20.5
25	2.40	20.0

Air-dry 2-g (0.07-oz) seed samples were placed in petri dishes suspended over saturated paper towels in closed chambers in full room shade (indirect light) and ambient July air temperatures (ca. 70-95°F; 21-35°C) Data are given as weight and as moisture uptake (per cent air-dry weight).

Dashed line represents break in curve at maximum moisture uptake. Sample I (in clear plastic chamber) was subjected to ambient light; sample II (in a light-proof metal chamber) received light only at the times of weighing.

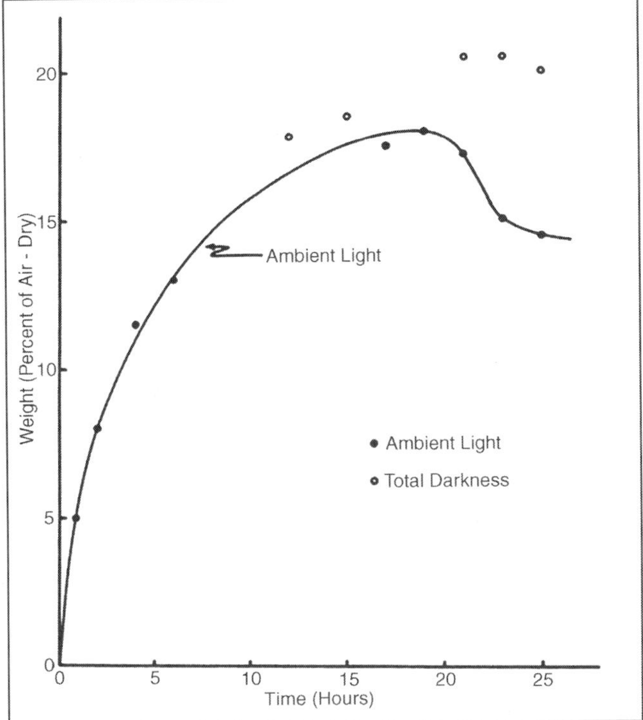

Fig. Hygroscopic Moisture Uptake by Saguaro Seeds at High Relative Humidities (90-100 per cent).

In a near-saturated atmosphere without condensation, saguaro seeds absorb atmospheric moisture up to approximately 20 per cent of their air-dry weight. Maximum hygroscopic imbibition is reached in approximately 20 hr, *i.e.*, ca. one day. Data in Table below. In an experiment to determine the effect of such treatment on the required period of contact with free water, treated seeds reached 50 per cent germination in 46.5 hr, 21.5 hr sooner than untreated (air-dry) seeds (68.0 hr).

Hygroscopic imbibition can reduce by approximately one day the required pre-germination period of seed contact with free water. Number and per cent germination of humidified (pre-treated) and of air-dry (untreated) saguaro seeds in the laboratory under ambient summer light and temperatures (full shade, ca. 65-95°F; 18.3-35.0°C) Pretreated seeds were held 5 days in open petri dishes in a near saturated atmosphere immediately prior to testing. Hours to germination of treated (2—1.00 seed lots) and untreated (3—10.0 seed lots) seeds is measured from time of initial contact with free water (15 August, 0800 hr). Data graphed in Figure.

Hygroscopic imbibition can reduce by approximately one day the required period of seed contact with free water to germinate under natural conditions. Thus, in the desert environment where moisture is critically limiting, chances for germination are greatly increased by this adaptive mechanism that can reduce—by as much as a whole day—the required period of contact with a saturated soil surface. Germination 50 per cent points determined by regression analysis after probit transformation.

Hours Wet	Germination			
	Humidified (20.0 Seeds)		Air Dry (300 Seeds)	
	No.	%	No.	%
24	7	3.5		
26	16	8.0		
28	27	13.5		
30	39	19.5		
42	95	47.5	11	3.7
48	116	58.0	22	7.3
54	148	74.0	88	29.3
72	167	83.5	169	56.3
120	184	92.0	254	84.0

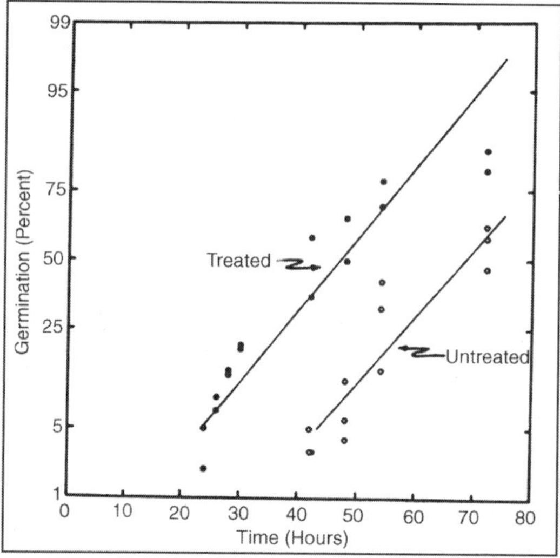

Fig. Effect of Pre-exposure to High Relative Humidity on the Rate of Saguaro Seed Germination after Contact with Free Water. The Time for Seeds Pre-exposed (Pretreated) to a Near-saturated Atmosphere to Reach 50 per cent Germination (46.5 hr) was Approximately 21.5 hr Less than the Time (68.0 hr) for Untreated (Air-dry) Seeds.

Germination under natural conditions, therefore, need not depend upon the rare occurrence of a continuously saturated soil surface over a prolonged period of time; but rather it can be sustained by a period of high relative humidity followed by a relatively brief period of saturation. Thus, hygroscopic imbibition allows rapid germination with intermittent high moisture levels commonly associated with the characteristic summer storm patterns of the

Sonoran Desert. Requisite environmental conditions for saguaro seed germination occur far beyond the bounds of the plant's natural distribution. Only at the western boundaries of its occurrence—in the lower Colorado River Valley—where summer rainfall is seldom adequate to promote germination, does the range of the species appear to be limited primarily by germination requirements. With that exception, therefore, the factors that limit its distribution and control populations along the margins of its range must act during the post-germination stage of the plant's development.

FACTORS AFFECTING SEED GERMINATION

Seed germination depends on both internal and external conditions. The most important external factors include temperature, water, oxygen and sometimes light or darkness. Various plants require different variables for successful seed germination, often this depends on the individual seed variety and is closely linked to the ecological conditions of a plant's natural habitat. For some seeds, their future germination response is affected by environmental conditions during seed formation; most often these responses are types of seed dormancy.

WATER

Water is required for germination. Mature seeds are often extremely dry and need to take in significant amounts of water, relative to the dry weight of the seed, before cellular metabolism and growth can resume.

Most seeds need enough water to moisten the seeds but not enough to soak them. The uptake of water by seeds is called imbibition, which leads to the swelling and the breaking of the seed coat. When seeds are formed, most plants store a food reserve with the seed, such as starch, proteins, or oils. This food reserve provides nourishment to the growing embryo. When the seed imbibes water, hydrolytic enzymes are activated which break down these stored food resources into metabolically useful chemicals. After the seedling emerges from the seed coat and starts growing roots and leaves, the seedling's food reserves are typically exhausted; at this point photosynthesis provides the energy needed for continued growth and the seedling now requires a continuous supply of water, nutrients, and light.

OXYGEN

Oxygen is required by the germinating seed for metabolism. Oxygen is used in aerobic respiration, the main source of the seedling's energy until it grows leaves. Oxygen is an atmospheric gas that is found in soil pore spaces; if a seed is buried too deeply within the soil or the soil is waterlogged, the seed can be oxygen starved. Some seeds have impermeable seed coats that prevent oxygen from entering the seed, causing a type of physical dormancy which is broken when the seed coat is worn away enough to allow gas exchange and water uptake from the environment.

TEMPERATURE

Temperature affects cellular metabolic and growth rates. Seeds from different species and even seeds from the same plant germinate over a wide range of temperatures. Seeds often have a temperature range within which they will germinate, and they will not do so above or below this range. Many seeds germinate at temperatures slightly above room–temperature 60–75 °F (16–24 °C), while others germinate just above freezing and others germinate only in response to alternations in temperature between warm and cool. Some seeds germinate when the soil is cool 28–40 °F (–2—4 °C), and some when the soil is warm 76–90 °F (24–32 °C).

Some seeds require exposure to cold temperatures (vernalization) to break dormancy. Seeds in a dormant state will not germinate even if conditions are favourable. Seeds that are dependent on temperature to end dormancy have a type of physiological dormancy. For example, seeds requiring the cold of winter are inhibited from germinating until they take in water in the fall and experience cooler temperatures.

Four degrees Celsius is cool enough to end dormancy for most cool dormant seeds, but some groups, especially within the family Ranunculaceae and others, need conditions cooler than –5 C. Some seeds will only germinate after hot temperatures during a forest fire which cracks their seed coats; this is a type of physical dormancy. Most common annual vegetables have optimal germination temperatures between 75–90 °F (24–32 °C), though many species (*e.g.* radishes or spinach) can germinate at significantly lower temperatures, as low as 40 °F (4 °C), thus allowing them to be grown from seed in cooler climates. Suboptimal temperatures lead to lower success rates and longer germination periods.

LIGHT OR DARKNESS

Light or darkness can be an environmental trigger for germination and is a type of physiological dormancy. Most seeds are not affected by light or darkness, but many seeds, including species found in forest settings, will not germinate until an opening in the canopy allows sufficient light for growth of the seedling.

Scarification mimics natural processes that weaken the seed coat before germination. In nature, some seeds require particular conditions to germinate, such as the heat of a fire (*e.g.*, many Australian native plants), or soaking in a body of water for a long period of time. Others need to be passed through an animal's digestive tract to weaken the seed coat enough to allow the seedling to emerge.

SEED GERMINATION AND DORMANCY

Some knowledge of the biology of seeds is essential to their proper handling. The use of seed for artificial regeneration makes possible a considerable degree of control over the conditions in which it is collected, processed, stored and treated, but the seed's inherent characteristics have been evolved as a result of millenium of adaptation to natural regeneration under local conditions.

Knowledge of flowering phenology enables the collector to select the timing and methods of seed harvesting most appropriate to the species, while handling, storage and pretreatment of seed will benefit from a knowledge of how seeds develop in nature.

This chapter gives a very brief and simplified account of seed biology in angiosperms and gymnosperms, but there is wide variation in the details of seed development between different genera.

IMPROVING SEED GERMINATION

When we want to produce large amounts of plants or simply when we want to start our gardens fast and get the most out of our purchases improving seed germination becomes a large priority. One of the largest concerns of world agriculture as well as the home grower is the decrease in germination time and increase in germination percentage since both of these factors can bring great benefits. Some seeds – especially some flowers and herbs – are often quite difficult to germinate and using certain techniques to increase the rate and speed in which they sprout has been the focus of a large amount of scientific research.

The use of priming to decrease germination time, especially what priming is, what types are available and which ones you can use to decrease the germination time of those very difficult seeds.

To understand the concept of priming we first need a good grasp at the general concept of seed germination. A seed is a dormant embryo which carries within it the potential for a new plant's life. The seed is alive, yet has a very slow metabolic rate due to the low mobility of substances within the embryo's cells. This low metabolic rate allows the seed to remain alive, yet survive extremely long periods of time (some seeds can survive even hundreds of years) before actually sprouting into new plants.

GERMINATION

Is the process in which we awaken the embryo – increases seed metabolism and toggles the massive reproduction that causes a new plant to grow.

The main mechanism that triggers this process is simply liquid water. When water gets into the embryo and hydrates its cells, it speeds up metabolism and allows the process of cell division and growth to rapidly increase. However it is not always this simple to start this process since several impairments – both chemical and physical – can exist for successful germination.

Priming is simply a process done prior to conventional seed germination which allows the inhibiting mechanism to be broken and the metabolic speed increase to begin. There are several types of priming that can be done. A seed can be submerged in simple water (hydropriming), it can be soaked in a solution of a simple salt (halopriming) or it can be set in a non-ionic solution with high osmotic pressure (osmopriming). It is not entirely well established why one technique might work better than another but certainly some species tend to respond much more efficiently to one or another.

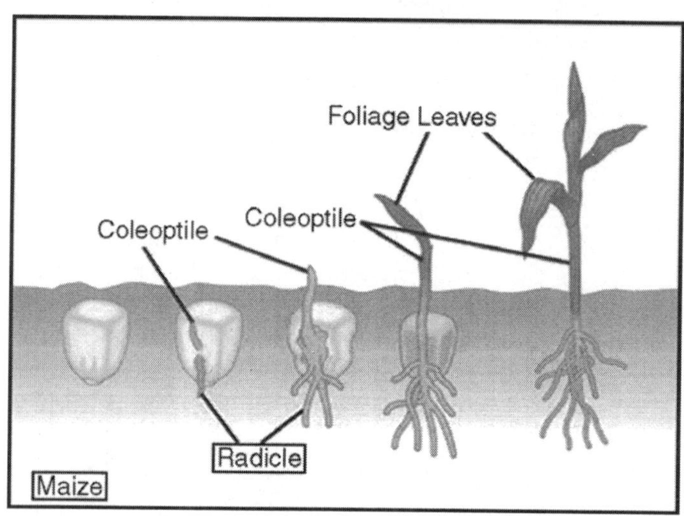

In general, priming offers the opportunity to almost always germinate seeds at much higher speeds without detrimental effects in germination percentages. For example, a two day treatment of parsley seeds with a PEG 6000 (PolyEthyleneGlycol) solution can reduce germination times substantially, from a few weeks to just a few days. Other seeds such as coriander might also benefit from similar treatments with PEG or with treatments with NaCl solutions. In general if you are looking to test priming on some difficult seeds you own you can try three small experiments to know which one works best for your particular seed variety and germination conditions.

Do one experiment in which the seeds are simply soaked in water for 24 hours, another in which seeds are placed in a 200mg/L NaCl solution and another one in which the plants are submerged in a PEG 6000 20 per cent solution, then let the seeds air-dry after the treatments. After comparing the results of these experiments with a control with no priming you will be able to see which priming technique is better for you and most effectively increases your seed germination rates.

To sum it up priming your seeds is a very efficient technique to increase the speed of germination without sacrificing germination rates. This methods are not very useful for seeds such as lettuce or tomato – which germinate easily – but they are invaluable for plants such as parsley, coriander or carrots which are generally much harder to germinate. If you have some seeds that have been giving you a hard time or seem to take ages to germinate then setting up some priming experiments might be the best thing to do. Teachers may be unfamiliar with the concept of "priming" seeds. However, teachers understand the value of Early Childhood Education ... and how it can influence later classroom experiences in the higher grades! Giving a child a good start in education is important; the same can be said for seeds. Priming seeds is like giving them a "better start" to life.

Another way of explaining this process has been developed by Keith Kubik, Ph.D. a Seed Physiologist with the Harris Moran Seed Company in California. Dr. Kubik compared seed priming to building a pre-fabricated house. Pre-fabricated houses are built Inside A Factory. The walls and roof are put together in an environment that is ideal for the workers – very important in our climate here in Canada - and building efficiencies are best. However, the house is not completed inside the factory.

The pre-built walls and roof are taken to the building site where the remaining construction takes place – the electrical work, the plumbing, and the basic fixtures like countertops and inside trim. Less time is consumed at the site, and the houses go up more quickly and more efficiently.

Similarly, with seeds, the priming process starts the germination process in the laboratory. The basic chemical reactions or framework needed for the seed to germinate take place under ideal temperatures and high moisture conditions.

The seed moisture is set so that the germination is ALMOST complete ... just like the prefabricated house when it is moved onto the actual home site. When you get your seeds, one of the two groups will be a control group and the other group will have been "primed" – almost ready for germination. Like the pre-fabricated house, seed germination of the primed seeds will probably take less time, because part of the germination process is already complete. You will still report on the germination process – the number of seeds planted and number of seeds germinated.

POLLINATION AND FERTILIZATION

A seed is a reproductive unit which develops from an ovule, usually after fertilization. Ovules are borne by both the angiosperms (true flowering plants) and the gymnosperms (which include the conifers). In the angiosperms the ovules are totally enclosed within the ovary, while in the gymnosperms the ovules are "naked", typically borne in pairs on the upper surface and near the base of each scale in a female cone.

Since the cone scales remain tightly closed except at the time of pollination and later at seed shed, the term "naked" is a relative one. Seed development is initiated by fertilization, the union of a haploid male nucleus from the pollen grain with a haploid female nucleus within the ovule to form a new diploid organism.

Fertilization must be preceded by pollination, the arrival of a pollen grain on the stigma of the female flower in angiosperms or close to the micropyle of the gymnosperm ovule. It is important to distinguish the two separate processes of pollination and fertilization (Fritsch and Salisbury 1947). In most angiosperms the elongation of the pollen tube is rapid and the interval between pollination and fertilization is only a few days or even hours. In a few angiosperms (*e.g.* Liquidambar, some species of Quercus) and many gymnosperms (*e.g.* Pseudotsuga, Larix, Picea) the interval is several weeks or months, while in other species of Quercus and in many Pinus it is a year to 14 months.

ANGIOSPERM SEED DEVELOPMENT

At the time of fertilization a typical angiosperm ovule consists of one or two protective coats—the integuments—and a central tissue—the nucellus. Often the integuments and the nucellus are clearly differentiated only in the region of the micropyle—the minute pore in the integuments through which, in many species, the pollen tube enters the nucellus. The ovule is attached to the wall of the ovary by a stalk—the funicle. Meiosis of a mother cell within the nucellus, followed by several mitotic cell divisions, leads to the formation of the embryo sac, a haploid eight-nucleate, seven-celled structure which occupies the central space within the nucellus. When the pollen tube reaches the embryo sac it releases two male gametes.

One male gamete unites with one of the nuclei in the embryo sac -the egg cell—to form a zygote which later develops into the diploid embryo plant. The second male gamete unites with two other female nuclei—the polar nuclei—to form a triploid cell which later develops into the endosperm, a tissue which acts as a food reserve for the growing embryo. The remaining five nuclei of the embryo sac (2 synergids and 3 antipodal cells) play no further role in seed development. Successful fertilization of the egg cell and successful triple fusion with the polar nuclei are both necessary for development of a viable seed.

Development of the fertilized ovule into the mature seed involves several different parts.

From the outside inwards these are as follows:

- The integuments of the ovule become the seedcoat of the mature seed. This sometimes consists of two distinct coverings, a typically firm outer seedcoat, the testa, and a generally thin, membranous inner coat, the tegmen. The testa protects the seed contents from drying out, mechanical injury, or attacks by fungi, bacteria and insects, until it is split at germination. But there is great variation in seedcoat among angiosperms.

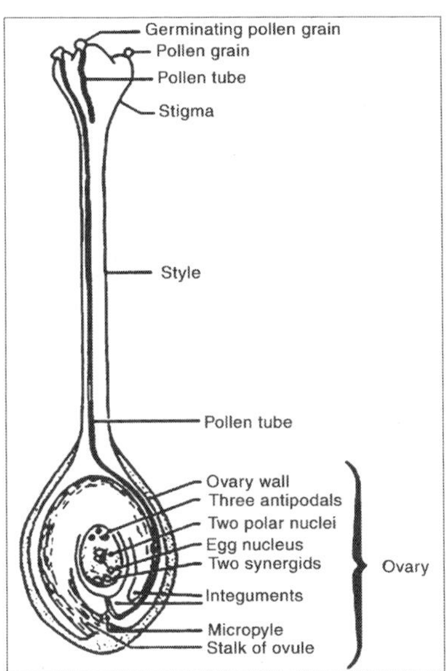

Fig. Longitudinal Section through a Typical Pistil Just Before Fertilization

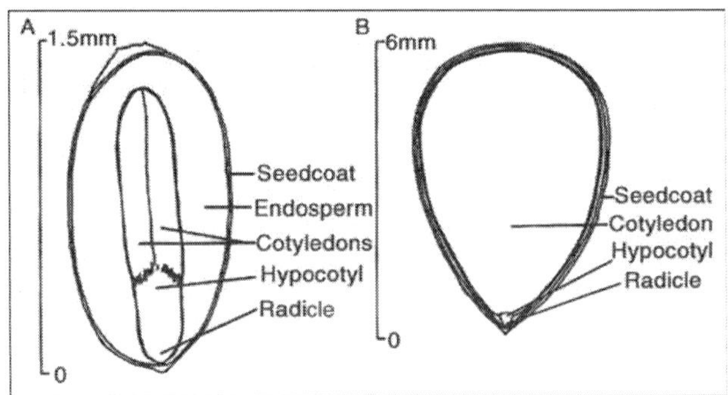

Fig. *Longitudinal Sections through Ripe Seeds of:* (A) *Paulownia TomenTosa:* ShowingConspicuous Endosperm. (B) *Tectona Grandis*: Endosperm has Disappeared and Cotyledons Occupy Almost the Entire Seed Cavity

Examples of different types of fruits:

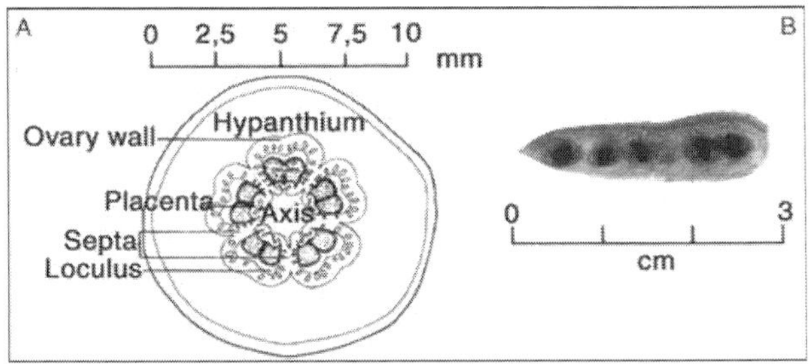

Fig. (A) Cross-Section of Capsule of *Eucalyptus Preissiana* Showing Loculi, Axis, Placentae and Ovules. (B) Open Pod with Seeds of *Acacia Aneura*

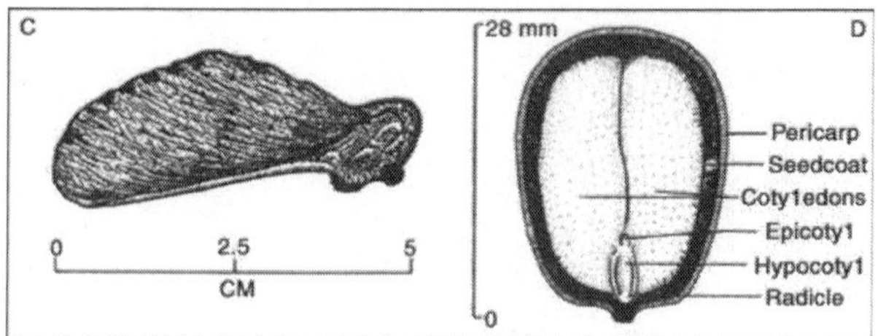

Fig. (C) Samara of *Triplochiton Soleroxylon*. (D) Nut (Acorn) of *Quercus Rubra*

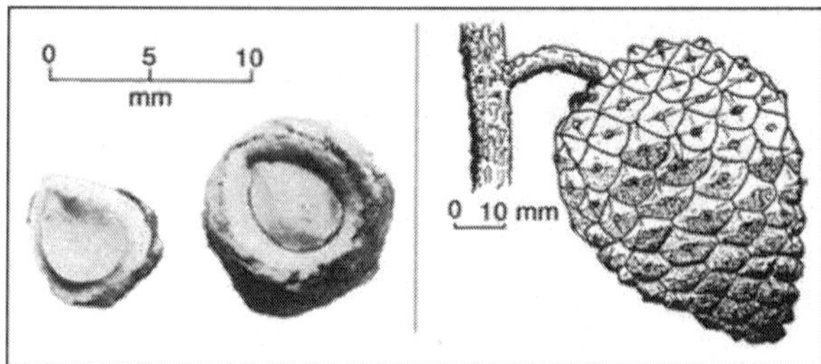

Fig. (E) Drupe of *Tectona grandis*. (F) Cone of *Pinus oocarpa*

- The nucellus may persist in some genera as a thin layer—the perisperm -lying inside the seedcoat and supplying food reserves to the embryo. In most angiosperms, however, it soon disappears and its function is taken over by the endosperm.
- The endosperm commonly grows more rapidly than the embryo during the period immediately after fertilization. It accumulates reserves of food and its fullest development is rich in carbohydrates, fats, proteins and growth hormones. In some species the endosperm remains conspicuous and still fills a greater part of the seed than the embryo even when the seed is ripe. In others, such as Tectona, the embryo absorbs food reserves from the endosperm during its later stages of development, until the endosperm disappears by the time the seed is mature.
- The embryo occupies the central part of the seed. Its degree of development at the time the seed is ripe varies greatly according to species. In some it is possible to distinguish all parts of the rudimentary plant—the radicle which at germination will give rise to the primary root, the seed leaves or cotyledons, the plumule from which will develop the primary shoot, and the hypocotyl which connects the cotyledons with the radicle. If the embryo absorbs all the food reserves from the endosperm, the thick fleshy cotyledons commonly become the main organs for food storage and occupy almost the whole of the seed cavity.

Although the storage function within the embryo is normally performed by the cotyledons, in Anisophyllea, Barringtonia and Garcinia it is completely taken over by the swollen hypocotyl which fills the seed cavity; the cotyledons are vestigial or absent. This is also the case in Lecythis and Bertholletia, hence the edible content of a Brazil nut is neither endosperm nor cotyledon but hypocotyl. In some species the embryo is still small and undeveloped when the seed is ready for dispersal, and an additional period under suitable environmental conditions is needed for maturation of the embryo after seed shed, before the seed can become capable of germination, *e.g.* Fraxinus excelsior.

At its most complex the ripe seed may thus consist of diploid tissue from the mother tree (the seedcoat, including testa and tegmen, and the perisperm), triploid tissue in the endosperm, and diploid tissue of the new genetic combination in the embryo offspring. But both perisperm (nearly always) and endosperm (not infrequently) may be missing. The essential constituents of all seeds are the embryo, the protective covering of the seedcoat and a reserve of food substances which may be stored in the cotyledons, hypocotyl, endosperm or perisperm.

Occasionally more than one embryo may develop in a single seed and such polyembryony has been reported from several genera. It is, however, the exception.

ANGIOSPERM FRUIT DEVELOPMENT

Development of the fertilized seed is normally accompanied by development of the fruit. In the simplest case the ovary wall becomes thickened to form the pericarp.

This may be:

- Dehiscent, splitting open when ripe to release the enclosed seeds; examples are the capsule (*e.g.* Eucalyptus), a multilocular fruit derived from a syncarpous ovary, and the leguminous pod (*e.g.* Cassia), which is derived from a single carpel and spilts along two sutures. The pericarp may be dry, semi-fleshy or fleshy at the time of dehiscence. Semi-fleshy to fleshy capsules are common in the humid tropics (*e.g.* Baccaurea, Durio, Dysoxylum, Myristica) and are often associated with the development of variously coloured, tasty or smelly pulp (aril or sarcotesta) around the seed.
- Indehiscent or dry, closely fused with the seed; examples are the achene, a small hard one-seeded fruit with membranous pericarp, thesamara, similar to the achene but with pericarp extended to form a wing (*e.g.* Triplochiton) and the nut, a rather large one-seeded fruit with woody or leathery pericarp (*e.g.* Shorea, Quercus).
- Indehiscent and fleshy, often distinguished by colour, smell and taste to attract fruit-eating birds and animals. Two types are distinguished. The berry has an outer skin and inner fleshy mass, containing seeds that have a hardened seedcoat (*e.g.* Diospyros, Pouteria). The drupe

has the inner layer of the pericarp hardened to protect the seeds (*e.g.* Prunus, Gmelina, Azadirachta, Mangifera); the seedcoat, having no protective function in a drupe, is usually papery or membraneous. The different pericarp layers in a typical drupe are known as exocarp (the skin), mesocarp (the flesh) and endocarp (the stone). The stone may be actually stony as in Gmelina or leathery as in Mangifera.

In some species other parts of the flower, as well as the ovary wall, take part in fruit formation. An example is the pome, found in apples and pears, in which the enlarged fleshy receptacle forms the greater part of the fruit, while the pericarp forms the core. An additional partial or entire protective covering may be provided by fused bracts arising below the flower—the involucre.

This may be papery, as in Tectona, or thicker and leathery as in the "acorn cup" of Quercus. Some fruits are formed by the coalescing of an entire inflorescence *e.g.* Morus, Chlorophora,Anthocephalus, Artocarpus. At the opposite extreme, in several genera of the Sterculiaceae, fruit formation does not occur at all in the normal angiospermous manner. Soon after fertilization, the carpel (follicle) splits on one side and develops into a large membraneous scale-like or boat-shaped wing; the fertilized ovule develops in a naked position at or near the base of the open carpel, in a gymnospermous manner. Such fruits must be the most primitive of all angiosperm fruits. At maturity the seeds are dispersed, attached to their carpels which now behave as wings.

The interval between flowering and maturation of seeds and fruits varies greatly with species, even within the same genus. In Eucalyptus it varies from one month in E. brachyandra to 10 – 16 months in E. diversicolor. In most Malaysian Dipterocarps it is between two and five months. In Tectona grandis it takes 50 days from flowering for the green fruits to develop to full size but 120 -- 200 days before they are fully ripe (Hedegart 1975).

In a study of rooted cuttings of Gmelina arborea in pots in Nigeria, individual flowers took 11 days from flower bud to opening and 45 days from flower bud to ripe fruits (Okoro 1978).

In Pterocarpus angolensis the interval between flowering and fruit maturation is 8 months (Boaler 1966). The shortest interval on record between flowering and seed maturation for a tropical timber species is apparently 3 weeks, for Pterocymbium javanicum. In contrast, some species of temperate Quercus take about 18 months from flowering to production of mature seeds.

In most species fertilization of one or more ovules must precede fruit formation. In a few species, however, fruits are set and mature without seed development and without fertilization of an egg. Such fruits, called parthenocarpic fruits, occur in several genera of forest trees including Acer, Ulmus, Fraxinus, Betula, Diospyros and Liriodendron. Mature fruits do not invariably indicate mature seed, still less can the number of sound seeds be predicted from the number of fruits. In Tectona the number of sound seeds per fruit can vary between 0 and 4 and still greater variation is possible in other genera.

SEED DISPERSAL IN ANGIOSPERMS

There is thus an immense variety among angiosperm fruits. Much of it is related to the need for seed dispersal. Survival and growth of young seedlings under the parent tree is often difficult, because of lack of light and intense root competition. Dispersal over a wide area can ensure that some seeds find conditions suitable for germination and survival, even though the vast majority will perish from the effects of harsh site conditions, competition or destruction by animals or disease.

Dispersal by wind is assisted when the seeds are very light and small *e.g.* Eucalyptus, or when either the seedcoat (Salix, Ceiba, Dyera) or the pericarp (Triplochiton, Pterocarpus, Koompassia, Casuarina, Fraxinus) possesses wings or hairs which serve to prolong flight. Fruits may also be winged by the enlargement of persistent sepals (most Dipterocarps) or persistent petals.

The distance of seed or fruit dispersal by wind depends not only on the weight and type of dispersal unit but also on the local wind conditions and the exposure and isolation of the mother trees. Studies of the winged fruits of Shorea contorta in the Philippines indicated that 90 per cent of the fruits travelled 20 m or less from the stem of the mother tree and a summary of other dipterocarp studies compiled by the same authors shows that most fruits landed within 30 m or, at most, 40 m. This compares with a dispersal distance within 2–3 m of the crown perimeter for a heavy, wingless seed such as Quercus crispula in Japan and a distance of over 60–90 m for 5 per cent of the light, winged seeds of Betula ermannii downwind from a belt of mother trees left in a logged over area.

Fleshy edible fruits and arillate seeds, on the other hand, encourage dispersal by birds or mammals. When such fruits or seeds are eaten by animals, the seeds, protected by the hard seedcoat or endocarp, often pass unharmed through the digestive tract and are deposited in the faeces at a considerable distance from the place where they were consumed. In many cases the digestive juices actually assist subsequent germination through softening of the hard seedcoat. In Africa the hornbill is a highly efficient dispersal agent for seeds of Maesopsis eminii. Sometimes the process is so effective as to be an embarrassment. Free-ranging goats eat the pods of Prosopis in some countries and spread the seeds indiscriminately over large areas; excellent germination and the aggressive pioneering qualities of the young plants may then render this genus a dangerous weed-tree. Coralling of the goats and collection of the seeds for use under strictly controlled management can overcome the problem. In other cases the fruit is eaten while the stones or seeds are rejected, but the animal may carry the fruit some distance from the parent before dropping the seeds. Rodents remove and store nuts or seeds; many are subsequently eaten but some may escape to germinate in the new situation.

Wind and animals are the most important agents of dispersal, but dispersal by water is common in some riverine species and large and heavy fruits are distributed to some extent by gravity on steep slopes.

GYMNOSPERM SEED DEVELOPMENT

Gymnosperm ovules have certain characteristics in common with angiosperm ovules, but there are a number of differences. There is normally a single protective integument which in a typical female cone is partially fused to the ovuliferous scale carrying the paired ovules. Within the integument is the nucellus which at fertilization, as in angiosperms, is clearly separated from the integument only in the region of the micropyle. Meiosis within the nucellus, followed by mitotic cell divisions, leads to the formation of a multicellular haploid tissue—the female gametophyte. By the time of fertilization it has developed much further than the 8-nucleate embryo sac in the angiosperms and has largely displaced the nucellus. At its micropylar end it is differentiated into one to many archegonia, each of which contains a large egg cell.

At fertilization the pollen tube releases two male nuclei into an archegonium, one of which unites with the egg nucleus. The resulting zygote later develops into the new diploid embryo. The second male nucleus aborts in Pinus but may fertilize a second archegonium in other genera *e.g.*Cupressus. It never unites with female polar nuclei to form a triploid tissue analogous to the endosperm of angiosperms; this type of tissue is unknown in gymnosperm seeds. For detailed descriptions of gymnosperm embryogeny, readers are referred to the specialized literature.

The mature seed consists of some or all of the following:

- The seedcoat or testa developed from the integument, diploid from the female parent.
- The diploid perisperm, developed from the nucellus. In most species this is absorbed by the female gametophyte and has disappeared by the time the seed is ripe, but it is still recognisable as a distinct tissue in *e.g.* Pinus pinea.
- The haploid female gametophytic tissue which serves as a food storage organ to nourish the embryo. Its function is the same as the endosperm in angiosperms and it is frequently called by that name, though this usage has been deprecated (Bonner 1984a).
- The embryo, with the same parts of radicle, cotyledons, plumule and hypocotyl as in angiosperms.

The number of cotyledons varies between and within genera, being up to 18 in Pinus, compared with the constant two in the dicotyledons which comprise the great majority of angiosperm trees. The essential constituents of embryo, protective covering and food storage tissue are present in all gymnosperm, as in all angiosperm, seeds.

More than one archegonium may be fertilized within a single ovule, but in the great majority of cases only one embryo per seed develops to maturity. Polyembryony does occur but is uncommon in most genera.

GYMNOSPERM FRUIT DEVELOPMENT

After fertilization the female cone which is typical of several important gymnosperm genera *e.g.* Pinus, Picea, Pseudotsuga, Araucariaincreases in size and weight, in moisture content and accumulated food reserves. As the cones approach maturity, the moisture content decreases again, accumulated food reserves move from cone to seed and the cone becomes more or less woody.

In Pinus a thin membranous flake becomes detached from the ovuliferous scale and adheres to the ripe seed, forming a wing. In Juniperus the cone scales grow together to form a fleshy berry-like fruit, while in Podocarpus and Taxus each singly borne seed becomes partly enclosed in a brightly coloured cup, the aril. The woody cone is, however, the most characteristic type of fruit in gymnosperms.

As in angiosperms, there is wide variation in the interval between flowering and seed maturity and dispersal. Because of the lengthy gap between pollination and fertilization in pines, mentioned earlier in this chapter, the total period between pollination and cone maturity is usually about two years in this genus; among tropical pines, average periods are 23 months in Pinus kesiya, 18–21 months in P. oocarpa. In Agathis robusta it takes 16 months from pollination to cone maturity, in Araucaria cunninghamii up to 24 months, in Araucaria hunsteinii 21–24 months. In several temperate genera development is completed within a single season *e.g.* in 5 months in Pseudotsuga menziesii.

Development of unpollinated female cones with fully formed but usually empty seeds occurs in a number of gymnosperm genera. Parthenocarpy is common in Abies, Juniperus, Larix, Picea, Taxus and Thuja. It is rare in pines.

SEED DISPERSAL IN GYMNOSPERMS

In the typical gymnosperm cone, ripening and drying of cone and seed causes the cone scales to open and release the seeds. Dispersal is by wind, assisted by the presence of seed wings in some genera *e.g.* Pinus. In some species of pine, the "closed-cone pines" *e.g.* P. radiata, there is usually an interval of months or years between ripening of cone and seed and the opening of the cone to release the seeds. In a few cases, such as the interior provenances of Pinus contorta, the cones open only when subjected to the heat of occasional fierce forest fires. On the other hand the cones of Abies and Araucaria disintegrate readily on the tree within a few weeks of ripening.

Seed dispersal by animals is less common, but the "berries" of Juniperus and the fleshy fruits of Podocarpus are examples. In addition seeds of temperate conifers are collected and stored by rodents and some may germinate before being eaten.

SEED GERMINATION

At one extreme, certain species of mangrove are viviparous, the seeds germinating before they separate from the parent. At the other extreme, seed of some species may remain dormant but alive for many years, capable of germinating if an event occurs to break the dormancy state. The subject of dormancy is discussed later in this chapter. Between the viviparous and the deeply dormant seed types occur many types of seed which are capable of germination soon after seed shed provided that environmental conditions are suitable. Just as fertilization initiates the transformation of the ovule into the ripe seed, so does germination transform the embryo within the seed into the independent seedling. For the purpose of laboratory testing, germination is defined as the emergence and development from the seed embryo of those essential structures which are indicative of the seed's capacity to produce a normal plant under favourable conditions (Justice 1972).

At maturity and seed shed many seeds have lost the greater part of the moisture which they contained in earlier stages. For example the embryo and female gametophyte of Pinus lambertiana contain as much as 50 per cent moisture content (fresh weight basis) shortly after fertilization, but by the time of natural seed dispersal the moisture content of the embryo is reduced to 23 per cent and of the female gametophyte to 38 per cent (Krugman *et al.* 1974). Reduced metabolic activity is associated with the drying of the seed, so that the embryo is in a temporarily resting or quiescent state, which in non-dormant seeds can be easily reactivated by suitable conditions.

These conditions are:

- Adequate moisture
- Favourable temperatures
- Adequate gas exchange and, for some species,
- Light.

There is considerable variation between species in the optimum levels of the different factors and there is frequently an interaction between them.

GERMINATION CONSISTS OF THREE OVERLAPPING PROCESSES

- Absorption of water mainly by imbibition, causing a swelling of the seed and eventual splitting of the seedcoat,
- Enzymatic activity and increased respiration and assimilation rates which signal the use of stored food and translocation to growing regions,
- Cell enlargement and divisions resulting in emergence of radicle and plumule.

In most seeds the radicle of the embryo is close to the micropyle, where absorption of water is easier and quicker than through the seedcoat. As the radicle swells, it exerts pressure on the seedcoat which commonly splits first at this point to free the radicle. This gives rise to the primary root which grows down into the soil and soon produces lateral roots. Subsequent stages depend on whether the species exhibits epigeal germination *e.g.* Pinus—the hypocotyl elongates and the cotyledons are lifted above ground—or hypogeal germination *e.g.* Quercus—the hypocotyl is undeveloped and the cotyledons remain on or in the ground. In hypogeal germination the cotyledons can have only a food storage function, or a haustorial function (in species in which food is stored in the endosperm *e.g.* palms, Scorodocarpus), while in epigeal germination they may also perform a valuable photosynthetic function during early growth of the seedling. In epigeal germination anchoring of the young plant by the radicle is followed by rapid elongation of the hypocotyl which arches upwards above the soil surface and then straightens; simultaneously the cotyledons and plumule are exposed, to which the seedcoat may or may not still be attached. The plumule then develops into the primary shoot and photosynthetic leaves. In the subtype "durian germination the hypocotyl elongates but the cotyledons are shed while still enclosed within the seedcoat (*e.g.* Durio zibethinus, Strombosia javanica). In hypogeal germination, the cotyledons remain in situ underground or on the ground while elongation takes place in the plumule. In the subtype "semihypogeal germination" the cotyledons are exposed but remain on the ground. The two main types, epigeal and hypogeal and the two subtypes, durian and semihypogeal are the result of the four possible combinations of two independent variables: hypocotyl elongated or not and cotyledons exposed or not. All four combinations occur in the humid tropics.

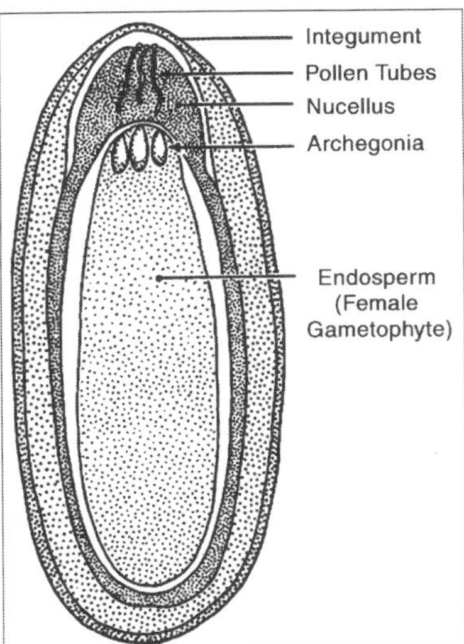

Fig. Longitudinal Section through an Ovule of *Pinus* During the Period of Pollen Tube Development Preceding Fertilization

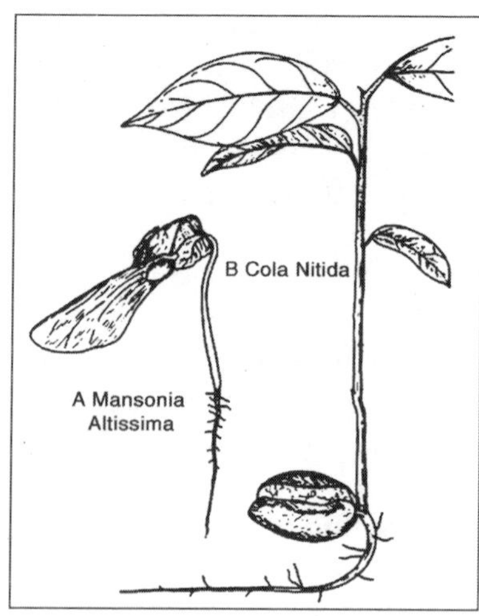

Fig. Examples of Germination in Two West African Sterculiaceae

Even in non-dormant seeds there is considerable variation, between species and individuals, in the speed of germination, from a few days to several weeks; much of this is due to different rates of imbibition in the first stage. Many tropical rain forest species, with high moisture content and permeable seedcoats at seed fall, must germinate within a few weeks. If they fail to find suitable conditions soon, they lose viability and die.

DORMANCY

The term "dormancy" refers to a condition in a viable seed which prevents it from germinating when supplied with the factors normally considered adequate for germination—suitable temperature, moisture and gaseous environment. A viable seed is defined as one which can germinate under favourable conditions, providing any dormancy that may be present is removed.

Dormancy in nature serves to protect the seed from conditions which are temporarily suitable for germination but which quickly revert to conditions too harsh for survival of the tender young seedling. Thus a seedcoat relatively impermeable to moisture prevents germination during isolated showers in the middle of a long dry season, while permitting it during a sustained rainy season. In the cool temperate zone the type of embryo dormancy which can be removed only through exposure to low temperatures facilitates subsequent germination in spring, while preventing it in autumn, when the resulting seedling would be unlikely to survive the winter. The strength of dormancy has been observed to vary according to latitude and provenance, and from year to year even in seed from the same parent. There is also differential dormancy within the same species and seedlot, so that germination is staggered over a more or less extended period of time. In the Malaysian woody flora, about 50 per cent of species complete germination within six weeks or less, so that the differential is not very great, but in some species such as the hard-seeded leguminous Parkia javanica the germination period may extend from 1 week after sowing for the first seed to two years for the last. Differential dormancy and staggered germination insures against the entire seed crop being destroyed by a single climatic catastrophe or pest.

In nature a number of external factors may work, more or less slowly, to end seedcoat dormancy. These include alternate heating and cooling, alternate wetting and drying, fire and the activities of animals, soil organisms, fungi, termites and other insects. Dormancy due to embryo immaturity will come to an end if the embryo is given time and suitable conditions in which to mature after seed shed.

The exact mechanisms of physiological dormancy of the embryo, and of the processes which can terminate it, have been widely investigated but underlying causes are still little understood. There is good evidence that growth-promoting hormones, of which gibberellin is a well known example, and growth-inhibiting hormones interact in the maintenance or breaking of dormancy. In temperate climates the balance between inhibitors and growth-promoters is altered by a combination of low temperature and high moisture maintained over a period of time which varies from species to species. This combination is provided naturally during winter, the season least suitable for growth. It can trigger biochemical changes in the embryo which lead to the breaking of dormancy, the initiation of embryo metabolism and growth, and the subsequent germination of the seedling.

Research on the physiology of tropical tree species has, unfortunately, been only a fraction of that done on temperate species. There is no reason to suppose that the combination of low temperature and high moisture—the technique of "stratification" when applied artificially—would have any effect on tropical seeds possessing embryo dormancy (if they exist). In the dry tropics, on the contrary, a combination of conditions typical of the season least favourable for growth—high temperature and low moisture—would seem more likely to trigger the breaking of embryo dormancy and lead to germination in the following rainy season. In fact, seedcoat dormancy by itself appears to provide adequate protection for species of the dry tropics.

From the forester's point of view dormancy has some disadvantages. Delayed and irregular germination in the nursery is a serious constraint on efficient nursery management (Bonner *et al.* 1974). Much research has therefore gone into devising effective artificial treatments to remove dormancy, in order to ensure that the seeds germinate quickly and evenly in the nursery beds. These treatments are described in Chapter.

On the other hand, dormancy confers certain advantages. Not only does it improve the chances of survival in nature, as mentioned previously, but it preserves the seed against temporarily unsuitable conditions such as may occur during the period between seed collection and storage. High quality orthodox but non-dormant seeds, dried to the appropriate moisture content and stored at the correct temperature, should have as long a life in storage as dormant seeds, but dormancy provides an insurance against the loss of viability during transport and processing which can easily occur in non-dormant seeds in less than ideal conditions.

SEED PRODUCTION OF HAZARDS

Both the quantity and the quality of seed crops may be greatly affected by external factors. Climatic factors can affect the abundance of flowering and thus indirectly of seed production. There is some evidence that above-average temperatures and a modest degree of moisture stress in spring and early summer can induce abundant formation of flower buds in temperate regions. In Nigeria good seed years occur in Triplochiton scleroxylon following a particularly dry (30 per cent or less of average rainfall) August, the month when there is a diminution of rainfall between the heavy early and heavy late rains.

More extreme cases of unseasonal climate usually reduce the crop of flowers or fruits. Late spring frosts in temperate regions kill flowers or young fruits and abnormally high temperatures or drought may have a similar effect. Even if death and premature shed of whole fruits does not occur, a proportion of the seeds may abort later. Mechanical destruction of flowers or fruits may be caused by exceptionally high winds or hailstorms.

Continuous rain at the time of pollen dispersal has a particularly adverse effect on the amount of seed set, whether pollination is by wind or insects. Tectona flowers during the rainy season and this may account for the low average rate of fertilization of 1–3 per cent reported from Thailand over the period 1967–72. Rain discourages flight of the pollinating insects as well as washing off pollen grains from the stigma before they germinate.

Continuous wet weather during the pollen dispersal season is considered the main factor responsible for the characteristically poor seed crops of Pinus merkusii in Indonesia and Malaysia. Tamar reported that over 90 per cent of Dipterocarp flowers in Malaysia failed to develop into fruits.

Birds, mammals, insects, fungi and bacteria all do damage in flowering as well as fruiting stages. Insects are probably responsible for the most serious losses in the greatest number of species. For example, the weevil Apion ghanaense destroys a large part of the flowers and seeds of Triplochiton each year. The larvae of Pagyda salvaris may destroy as much as 90 per cent of the flower buds of Tectona in some years. Two species of the Bruchid genus Amblycerus can destroy many seeds of Cordia alliodora, but damage can be reduced by collecting the seeds three weeks before natural seedfall.

The weevil Nanophyes sp. may attack up to 60 per cent of the seeds ofTerminalia ivorensis. Cone worms of the genus Dioryctria have been known to damage 60 per cent of maturing cones and seeds of Pinus elliottii and P. palustris in the southern USA and the same genus can also cause severe damage on Pinus merkusii seeds in the Philippines. Larvae of Agathiphaga, a genus of moths, may destroy over 50 per cent of the seeds in cones of several species of Agathis in Queensland and the western Pacific islands. Seeds of many dry area species of Acacia and Prosopis suffer serious damage from Bruchid larvae. Birds and mammals, especially squirrels, may consume considerable numbers of seeds in some years, although they also perform a useful service in seed dispersal. Losses from pests and diseases do not normally have a serious effect in years of abundant seed production, but in years when flowering is poor for climatic reasons they can convert a light crop into complete failure.

4

Germination Testing of Seeds

Of all the quality measurements of seed lots, none is more important than the potential germination of the seeds. The main aim of a laboratory germination test is to estimate the maximum number of seeds which can germinate in optimum conditions.

The use of standardized ideal conditions in the laboratory such as those prescribed by ISTA ensures that results obtained from a given seed lot in one laboratory should be identical with those obtained from any other laboratory in the same or in other countries. A common standard for assessing germinative potential is very valuable for seed moving in international trade. On the other hand, it is clear that results obtained under ideal controlled conditions in the laboratory are not directly applicable in the field nursery, where only limited control over environmental conditions is possible. Each nurseryman should apply his own correction factor, derived from experience over the years, to convert the germination potential of a seed lot, as determined from laboratory tests, to the actual field germination he can expect under his own local conditions.

At the other extreme, a nurseryman may prefer to carry out his own germination test in his own nursery in advance of large scale sowing. The results of the test should be directly applicable to the later sowings of the same seed lot in the same nursery, but would not be applicable to other nurseries.

However, there is sometimes insufficient time for germination tests in advance of the main sowing and managers of large operational nurseries may be reluctant to involve themselves in small scale research.

Between the two extremes lies the case of the small silvicultural research station which lacks the laboratory facilities to comply with ISTA prescriptions for testing, but which carries out germination tests in the nursery on bulk seed supplies before distributing them to afforestation projects throughout the country.

Here again the local forester or nurseryman must apply his own correction factor to convert actual germination figures from the research nursery to expected germination figures in his large operational nursery.

The following account is largely based on that of Turnbull. It is in conformity with ISTA rules (1976), but some of the principles are equally applicable if lack of equipment or trained staff makes it necessary to use simpler methods. Germination is defined as the emergence and development from the seed embryo of those essential structures which are indicative of the seed's capacity to produce a normal plant under favourable conditions (Justice 1972, ISTA 1976). Germination is expressed as the percentage of pure seeds which produces normal seedlings or as the number of seeds germinating per unit weight of the sample.

In the laboratory the environmental conditions, including moisture, temperature, aeration and light, must not only be specific enough to initiate germination but also favourable for the development of the seedlings to a stage where interpretation of normal and abnormal types may be made.

With a few exceptions, all germination tests should be made with pure seeds separated by the purity test. The pure seed must be well mixed and counted at random into replicates. The seeds should then be spaced uniformly on the test substrate. Normally one test consists of 400 seeds in 4 replicates of 100 seeds each, but if 100 seeds overcrowd the test substrate, the replicates may be broken down into a larger number of smaller replicates of 50 or 25 seeds each. A general recommendation is to leave 1.5 to 5 times the normal seed width or diameter between seeds, in order to discourage the spread of fungal moulds.

The exceptions are very small-seeded species in which it is impossible (some species of Eucalyptus) or very difficult (Alnus, Betula, Populus,Salix) to separate the seeds from the accompanying inert matter or chaff. In these cases the test is made with the same number of replicates but the replicates are of equal weight rather than an equal number of seeds.

GERMINATION EQUIPMENT

The kind of germinator can be selected according to the type and amount of seeds being tested and it is acceptable providing it can give good control of prescribed temperature, moisture, and light conditions.

Germinators range in size from small or portable individual units through cabinets of various sizes and large germination tables of the Jacobsen Copenhagen type, to walk-in germination rooms. The major types recommended by ISTA are as follows:

Jacobsen and Rodewald Apparatus

In Europe, a germinator called the Jacobsen apparatus or Copenhagen tank is commonly used. It consists of a bath containing water, the temperature of which may be controlled by thermostats. The seeds are distributed on top of paper and placed on metal or glass strips suspended about 5–7 cm above the bath; paper or cotton wicks lie beneath the substrate and pass through slots into the water below. The moisture content of the substrate may be adjusted by altering the water level, and so increasing or decreasing the distance between the seed bed and the water.

High humidity is maintained around the seed either by covering the entire tank with a transparent lid or placing an inverted plastic funnel with a hole in the conical end over each germination pad. The apparatus may be exposed to natural daylight but artificial light is usually desirable.

One of the disadvantages of the standard Copenhagen tank is the absence of direct temperature control at the seed bed. In the tropics it tends to overheat. Improved models have been developed, such as the one by Overaa which uses hollow stainless-steel strips, through which the water for heating and cooling is circulated. Other disadvantages are that the equipment requires a large amount of floor space for the number of tests accommodated. Also, the handling of covers and substrata appears to be more cumbersome than moving a lightweight tray with paper substrate.

The Rodewald apparatus consists of a zinc box covered with glass in which the seeds are exposed to direct or diffuse light. The bottom of the apparatus contains water over which a layer of moist sand is placed on a tray. The seed bed consists of unglazed porcelain dishes placed in the moist sand, or the dishes may be placed directly in the water. The temperature of the water is controlled thermostatically and the sand is moistened through wicks reaching into the water bath. A sand substrate is not desirable for species requiring alternating temperatures because of its slow adjustment to a change in temperature.

Germination Cabinet

Another common type of apparatus is the closed chamber for germinating seeds in darkness, diffused light, or direct light. This type of germinator often consists of a double walled cabinet, suitably insulated against temperature changes by an air jacket or layer of insulating material. It is equipped with appropriate tray slides to accomodate the type of germination trays preferred by individual laboratories.

The modern germination chamber has provision for both heating and cooling. Usually, water is chilled and passed between the walls of the chamber or through cooling pipes placed along the inner walls of the chamber. Either the air or a reservoir of water in the bottom of the chamber may be heated. In such cabinets the temperature can be regulated at any desired setting between approximately 8°C and 40°C.

Inexpensive germination cabinets are sometimes constructed locally. Gupta and Kumar (1977) describe the low cost germinators in use in the Seed Testing Laboratory in Dehra Dun, India. The outer wall of the cabinet is constructed of teak, the size 195 × 70 × 40 cm, and the electric controls include a thermostat, hot air blower, a separate air circulator and a time switch to control illumination. Glass wool is used for insulation between the inner and outer walls. Satisfactory performance is obtained at temperatures varying from room temperature up to 45° C (± 1° C).

A germination cabinet should meet as far as possible the following requirements (Oomen and Koppe 1969):

- *Air humidity*: As high as possible but preferably never lower than 90 per cent, to prevent the seed substrate from drying up;
- *Air temperature*: Adjustable between 10°C and 35°C, the temperature all over the working section of the cabinet over a number of days to be uniform in each of the temperature regimes, within plus or minus 1°C;
- *light*: Uniform illumination of the trays, intensity 750 to 1250 lux at the level of the seed;
- *Air movement*: As low as possible to prevent the seeds from drying up;
- *Fresh air supply*: small, of the order of one air change per hour, which should be sufficient to remove the carbon dioxide produced by the germinating seeds;
- Day and night cycling: A rapid initial temperature drop to be achieved in 30 minutes during the day to night change, reaching the ultimate night temperature in one hour, a similar requirement to hold for the night to day change;
- *Condensation*: Not allowed.

However, when cabinets operate with alternating temperatures it is very difficult to maintain the high humidity and the substrates dry out; this necessitates frequent watering, which adversely affects the germination results and adds to the manhours spent on the test. There are many modifications of the germination cabinet and for detailed descriptions of some of them see Justice (1972) and Oomen and Koppe.

Room Germinator

When large numbers of tests are made, whole rooms may be used as germinators, with temperature, humidity and light controlled.

The room germinator is a modification of the cabinet. It is constructed on the same principle as the cabinet, but is large enough to permit workers to enter it and place the tests along either side of a central passageway. Usually, fans are required to prevent stratification of temperature and special equipment is necessary to maintain a high relative humidity. Another modification is the combination room-cabinet germinator. The temperature of the entire room is kept as low as the lowest temperature required. Germination cabinets are placed in this room and heated, individually, by electricity to maintain the various temperatures required. Either constant or alternating temperatures may be obtained by this type of germinator.

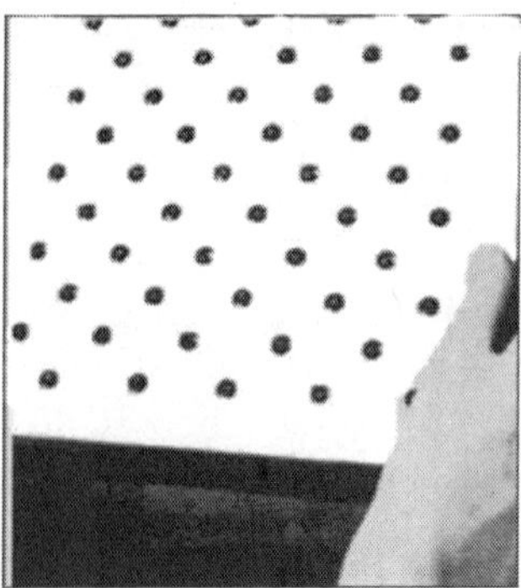

Fig. Counting board with seeds of *Celtis laevigata*. Seeds are spread on the top board to place one seed in each hole. The spring-held top board is then moved to the right until its holes line up with another set in the bottom board; the seeds then drop through.

Fig. Counting head on a vacuum seed counter. A pump pulls a vacuum through the line and the hollow counting head that has 50 or 100 tiny holes on its bottom surface. When one seed per hole is attached, the vacuum can be released with the finger plunger to drop the seeds.

Fig. Seed germination equipment at the Division of Forest Research, CSIRO Canberra (A) Open seed germination cabinet (B) Group of cabinets.

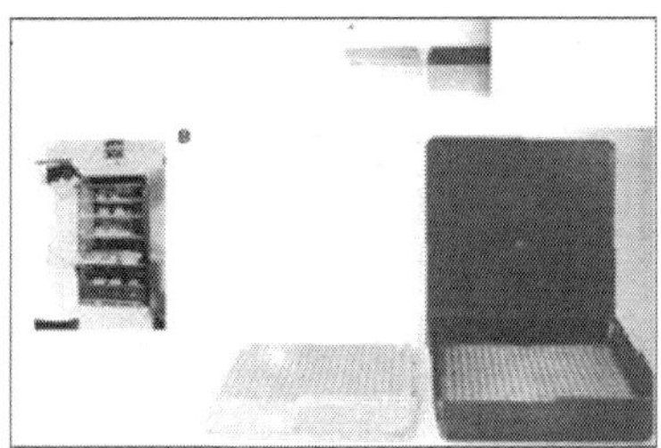

Fig. Conviron G30 germinator, with over 95 per cent relative humidity and programmable temperature and photoperiod within 24 hours, as used at Petawawa National Forestry Institute, Canada. (Canadian Forestry Service) A. Clear and black germination boxes developed for seed testing

Fig Copenhagen tank and rolled filter paper for germination tests.

Portable germination boxes. A simple and versatile germinator consists of a number of transparent plastic boxes with lids, which can be stacked one upon another.

Robbins (1984) describes the ideal container as being:

1. Rectangular and stackable, for economony of space;
2. Large enough for adequate spacing of at least one replicate of seeds (100, 50 or 25 seeds, depending on seed size);
3. Sufficiently deep to allow for the required substrate depth and to permit development of the seedling for proper assessment;
4. Provided with a wellfitting lid, to maintain a high moisture content of the substrate and surrounding air;
5. Easy to sterilize by heat or chemical treatment;
6. Transparent (at least the lid), to allow light if required for germination and subsequent development of the seedlings.

In Honduras the size of box used is 178 × 117 × 72 mm deep and each box contains one replicate of 100 pine seeds. This leaves a space of at least the width of one seed between each seed. For larger-seeded species there are fewer seeds per replicate box. Any of the common substrates described, *e.g.* filter paper, sand, may be used. By adding a suitable quantity of water to the substrate at the beginning of the test and then keeping the boxes covered with their lids at all times except during assessment and removal of seedlings, it is possible to maintain a high and constant moisture content in the substrate and the air inside the boxes, without the need for further additions of water or for control of humidity in the atmosphere outside the boxes. Where filter paper or blotting paper is used as the substrate, it can be kept permanently moist by placing it on a raised platform over a reservoir of water which is fed to the substrate by wicks. This arrangement resembles, in miniature, the Jacobsen tank. The boxes may be placed inside an incubator where temperature and light can be controlled in accordance with *e.g.* ISTA recommendations or, if no incubator is available, they can be stored in ambient room conditions of temperature and light. In either case the closed boxes should provide optimum humidity conditions for the germinating seeds.

Researchers at the Petawawa National Forestry Institute in Canada have developed a germination box, moulded out of light, unbreakable and heat-resistant polycarbonate plastic, which is large enough to contain 4 replicates of 100 seeds each of pines or similarly sized seeds. The box is 28 cm long × 24 cm wide. The base is 5 cm deep and the lid 1 cm deep, but the special universal locking mechanism makes it possible to fit two bases together; alternative combined depths of 6 and 10 cm are thus available for use according to the characteristics of the species under test. A perforated false bottom on eight legs 1 cm long supports the germination substrate and the space below can be used as a water reservoir if desired. The material is available in either clear or black plastic for germination in light or dark conditions; to ensure complete darkness it is necessary to seal the junction between base and lid with opaque tape. Four aeration holes on the side wall of the bottom component are optional in the clear plastic type. Tests of this germination box gave results closely comparable with those from the much smaller, and thus less convenient, petri dishes. It was found that loss of moisture over a four week period was least in boxes which had the water reservoir filled at the start of the test and which had no aeration holes. Both clear and black were equally good in this respect, but growth of mould was more severe in the (sealed) black than in the clear model. A substrate of blotting paper over Kimpak (cellulose paper) lost moisture more slowly than Kimpak alone.

Selection of a germinator. Most forest seed laboratories with a moderate through-put of samples use either the cabinet-type germinators or germination tables. A report by Boeke et al. (1969) recommends the exclusive use of cabinettype germinators for small seed testing laboratories. The advantages of using cabinets instead of germination tables of the Copenhagen type are a saving in space and, in well constructed cabinets, the more precise control of temperature and humidity (Oomen and Koppe 1969). It should be noted that a single-level Copenhagen tank requires five to eight times the floor space occupied by a cabinet of equal capacity (Boeke et al. 1969). Where research is being carried out on germination conditions, or where testing covers a wide variety of seeds, the flexibility provided by several cabinets operating at different temperatures and/or light conditions can be very advantageous. For nurseries or small research institutes concerned with carrying out simple tests of germination without maintaining strict control of temperature and light, the flexibility, storability and cheapness of plastic germination boxes have much to recommend them. They can also be used in combination with incubators in which light and temperature are controlled.

GERMINATION CONDITIONS

The optimum conditions for different stages of germination and seedling growth are not identical and may even vary for different seeds within the same seed lot. The aim of much seed testing research has therefore been to determine a combination of conditions which will give the most regular, rapid and complete germination for the majority of the same species.

Substrate

Soil is rarely used as a substrate for germination tests because each sample will vary greatly in physical, chemical and biological properties. While the germination results might be more comparable with field conditions, the lack of reproducibility and difficulty in comparing tests of different seed lots render it unsuitable. Artificial media are much more easily standardized. Most laboratory tests on small seeded species are made on paper. Other materials used include sand, granulated peat moss and expanded mica (Vermiculite and Terralite).

The main requirements for the substrate are:

- Non-toxic to the germinating seedlings
- Free of fungi and other micro-organisms
- Of porous texture to enable adequate aeration and moisture for the germinating seeds.

The choice of media on which the seeds are placed for germination depends on the equipment, the species, the working conditions and the experience of the operator. It is prescribed in the ISTA Rules for a number of forest trees. Filter paper, heavy box paper, or any other absorbent paper may be linked with a filter paper or cotton wick dipped into water, or placed on top of a layer of sand or vermiculite. Cellulose paper is becoming more commonly used as a germination medium because it is easier to handle than sand, but still permits penetration of the radicle, thus providing for better evaluation of abnormal germination. Also moisture content does not tend to become stratified within the paper medium as it does with sand.

The best paper substrates are germination blotters, paper towelling, laboratory filter paper and creped cellulose paper wadding. Any paper substrates should be checked for presence of toxic chemicals. The 1976 ISTA rules give detailed specifications for paper and towels, including weight, bursting strength, capillary rise and acidity. Seeds without a specific light requirement may be placed on top of paper or between folded paper. Folded paper increases the surface contact area between the seeds and the moisture supply. Large seeds may be rolled in paper towels which are then placed in an upright position; this allows the roots to grow downwards and avoids tangling of the roots. Rolled or folded paper packs may be kept moist without the necessity of a wick dipping into water. Rolled seeds may be stored on glass plates above, but not in contact with, a water bath and covered by a polythene sheet. Any rolls which show signs of drying out may be sprinkled with water. Use of rolled paper is a very rapid and convenient method but may cause curling of the radicles. This is of no concern if the germinated seeds are to be discarded after testing. For operational germination of large quantities of seed, however, the slightly slower folded paper method is preferable. This is used with success as standard practice for Pinus and Eucalyptus spp. in Thailand. Sand is not suitable for very small seeds because they are difficult to locate, but it is widely used for larger seeds. Sand can be sterilized and fungi develop on it less freely than on paper. It also provides a good contact between the seed and moisture because the seeds can be pressed into the medium. A commonsense rule is to cover the seed with a thickness of sand not less than the length of the seed on its longest axis. ISTA recommends a thickness of 1 – 2 cm according to seed size and prescribes that sand particle sizes should be between 0.05 and 0.8 mm and pH between 6.0 and 7.5. Sand as a germination substrate is preferred for tree species that have a longer germinative period, for instance, Rosa spp., Pinus caribaea, P. elliottii, and P. palustris, or with larger seeds, for example, P. pinea and Quercus spp. A sand-perlite medium permits leaching of the water which can be important when testing repellent-coated seed, but may be a disadvantage when testing seeds which are sensitive to drying.

Moisture and Aeration

The moisture level of the substrate has been suggested as one of the major causes of variation in seed research results. The ISTA Rules specify that a sand substrate be moistened depending on its characteristics and the size of the seed to be tested and suggest that a moisture level of 50 – 60 per cent of the water-holding capacity of the sand is suitable for several groups of agricultural seeds. A paper substrate must not be so wet that a film of water forms around the seed. Belcher, working with Pinus spp., concluded that most species have considerable tolerance to diverse moisture levels and the major variations in germination were due to dry conditions. He found some species to be sensitive to dry conditions and others to very wet conditions. As a generalization, the substrate must be moist enough at all times to supply the necessary moisture, but excessive moisture will restrict aeration. All tests must be examined daily to ensure that the moisture content of the substrate is near optimum. The water should be reasonably free of impurities.

Temperature Control

In laboratory seed testing the important temperature is that at the level of the seed. The temperature will vary for different species and appropriate values are listed for a large number of tree species in the International Rules for Seed Testing. In addition to temperature, it indicates prescribed substrates, light, counting periods and pretreatment. Temperature is one of the most critical factors in the laboratory germination of seeds and must be regularly checked. When alternating temperatures are required the test is usually held at the low temperature for 16 hours and at the high temperature for 8 hours each day. Temperature of 20° and 30°C alternating are commonly prescribed for tree species. Although natural fluctuations between day and night temperatures are less in lowland moist tropical forests than in other forest types, alternating temperatures may still affect germination of tropical species. In one experiment of Terminalia ivorensis in Nigeria, alternating temperatures of 34° C and 24° C gave 93 per cent germination in 41 days compared with 27 per cent at a constant temperature of 30° C, both with continuous light.

Light

Light is required for germination in many tree seeds. Fluorescent light is as effective as natural daylight and is preferred in testing, because the wavelength and intensity can be standardized, within limits. Cool white fluorescent lamps are recommended because of their light quality and low heat emission: Light should be evenly distributed over the tests, with an intensity within the range 750 – 1250 lux. Seeds should be subjected to light for only part of the test period, 8 hours in every 24 hours is usual but a longer or shorter period may be beneficial to seeds of some species.

Cotrolling Fungi in Germination Tests

Laboratory practices that will minimize the spread of fungi include proper spacing of seeds, control of temperature, removing decayed seeds, proper aeration, and keeping the substrate only just wet enough to permit germination. Sterilization of laboratory equipment and the periodic disinfection of germination cabinets and other apparatus will assist. As a general rule, seed is not disinfected because seeds which decay are usually of poor quality. Magini (1962) states that a number of experiments carried out with different methods of disinfecting the seed just before or during germination have shown no reliable improvement in germination percentage. A small quantity of fungicide added to the water used to moisten paper rolls has, however, been found beneficial in testing conifer seeds in Denmark. In Australia spraying of Karathane fungicide at a concentration of 0.8 g in one litre of distilled water has also proved effective when testing eucalypt seed on moist filter paper over vermiculite.

GERMINATION CONDITIONS FOR SELECTED SPECIES

Conditions prescribed by CSIRO and ISTA for germination testing of selected tropical, sub-tropical and temperate species are tabulated in Tables 9.1, page 213 and 9.2, page 215 respectively.

Notes for table:

1. Temperature recommendations separated by a semi-colon indicate that the (constant) temperatures have been found satisfactory. Alternating temperatures are not necessary.

2. Unless otherwise specified, normal substrate used for testing eucalypts by Division of Forest Research, CSIRO, Canberra is moist filter paper over vermiculite in a petri dish. Where "vermiculite only" is recommended, the filter paper should be omitted.
3. A temperature shown in parenthesis *e.g.* (25) indicates that satisfactory results have been obtained with that temperature but a full range of temperatures has not been tested.

Species	Mean No. of Viable Seeds per g of Seed and Chaff	Seed Testing Recommendations				Special Recommendations 2)
		Weight of Replication (g)	Temperature (°C) 1)	First Count (days)	Final Count (days)	
E. brassiana	340	0.15	25	7	14	
E. camaldulensis	670	0.10	30	5	10	Light essential
E. citriodora	110	0.50	25;30	5	14	Vermiculite only
E. doeziana	130	0.40	25	7	28	Vermiculite only
E. deglupta	4,000	0.01	35	5	14	Vermiculite only
E. globulus subsp. globulus	75	0.70	25	5	14	
E. grandis	650	0.10	25	5	14	
E. microtheca	380	0.15	35	3	14	Light essential, Vermiculite only or stratify 3 weeks, then 20°C
E. regnans	180	0.30	15	10	21	
E. saligna	540	0.10	25	5	14	
E. tereticornis	600	0.10	25;30;35	5	14	
E. urophylla	460	0.10	(25) 3)	5	14	

This table indicates permissible substrates, temperatures, duration and additional directions, including recommended special treatments for dormant samples. For species in section 1, the methods in columns 2 – 6 are prescriptive and no others may be used. Where two figures are shown under temperature, *e.g.* 20 – 30, this indicates daily alternation of temperatures. The lower temperature should be maintained for 16 hours and the higher for 8 hours, in each 24 hour period. For species in section 2, other methods may be used provided that the method is indicated on the International Analysis Certificate. For certain species indicated in column 7, duplicate tests (with and without prechilling) are necessary; the germination phase of these should run concurrently. The less desirable methods are placed in brackets in the Table.

The abbreviations have the following meanings:

- TP - top of paper
- BP - between paper (including rolled towels and pleated paper)
- S - sand
- TS - top of sand
- L - light essential
- TT - topographical tetrazolium test
- 1) - reference to weighed replicates in Eucalyptus and other genera has been deleted by a 1981 ISTA amendment. A practicable and up-to-date guideline to testing eucalypts is contained in appendix 3 of Boland et al. 1980. Data from that appendix for a few of the more important species is reproduced in Table.

	Species	Prescriptions for:					
		Substrata	Temperature (°C)	Light	First Count (days)	Final Count (days)	Additional directions including recommendations for breaking dormancy
	1	2	3	4	5	6	7
	Secrtion 1 (Prescriptive)						
	Acacia spp.	TP	20–30(20)	L	7	21	(1) Pierce seed, chio or file off fragment of testa at cotyledon end and soak 3 hrs. or (2) (Soak seed 1 hr. in concentrated H_2SO_4, wash seed thoroughly in running water after acid treatment).
	Ailanthus altissima	TP	20–30	-	7	21	Removal of pericarp after soaking for 24 hrs. may speed up germination.
1)	Alnus spp.	TP	20–30	L	7	21	(May be tested also by using 4 weighed replicates of 0.10 to 0.25 gm each, according to species)
	Cedrela spp.	TP	20–30	L	7	28	
	Cryptomeria japonica	TP	20–30	L	7	28	
	Cupressus sempervirens	TP	20	L	7	28	
1)	Eucalyptus camaloulensis	TP	30	L	3	14	Use 4 weighed replicates of 0.10 gm each
	Liquidambar styraciflua	TP	20–30	L	7	21	Sensitive to drying in test
	Nothofagus obliqus	TP	20–30	L	7	28	No prechill and prechill 28 days at 3–5 ° C. Double tests.
	Pinus caribaea	TP	20–30	L	7	21	
	P. elliottii	TP	22;20–30	L	7	28	
	P. pinaster	TP	20	L	7	35	(1) No prechill and prechill 28 days at 3–5 ° C. Light for no more than 16 hrs. per day. Double tests. (2) (Use TT)
	P. radiata	TP	20	L	7	28	
	P. taeda	TP	22;20–30	L	7	28	
	Quercus spp.	TS(S)	20	-	7	28	Soak seed for up to 48 hrs. and cut off 1/3 at scar end of seed and remove testa.
	Robinia pseudoacacia	TP	20–30	L	7	14	(1) Pierce seed or chip or file off fragment of testa at

GERMINATION TESTING IN THE NURSERY

A good example of a suitable method for testing germination in the nursery is the procedure recommended for tropical pines in West Malaysia:

"Take 400 pure seed and divide into 4 samples each of 100 seed. Take 4 wooden or plastic boxes (opaque flexible plastic sandwich boxes are very good) of dimensions 30 × 30 cm and of depth about 10 cm and fill them slightly overfull with sieved sand to the top of the sides. Wet the sand by applying 0.5 litres of water to each box. Level the sand across the top of each box and make drills at 2.5 cm intervals and of depth 6 mm (for P. oocarpa the drills should only be 3 mm deep). Sow the seed (100) at intervals of 2.5 cm and gently press them on to the sand in the drill. Cover to a depth of 6 mm (3 mm for P. oocarpa) with sand which has been passed through a Mesh No. 12 (12/64" or 4.76 mm) but retained on a Mesh No. 8 (Size 8/64" or 3.18 mm). Lightly water with a fine mist spray. Cover with transparent polythene sheeting of 0.2 mm thickness mounted on the upper side of a square tight-fitting wooden frame of 5 cm thickness. Within 24 hours condensation of moisture will appear on the inside of the polythene cover. If moisture disappears at any time during the ensuing 7 days then give a light mist spray and close lid once more. Germination will usually begin by the 7th day and the cover can be removed. Keep sand moist. A seed is recorded as having germinated when it has reached a height of 1 cm complete with seed coat enclosing the cotyledons. Record all germination from 7th to 28th day separately for each of the four boxes. Remove seedlings as soon as they are recorded plus any diseased ones which may not yet have reached 1 cm height as they may infect others.

On the 28th day after sowing sieve the top layer of sand in each box through a No. 12 mesh screen (mesh size 12/64" or 4.76 mm) and record the number of full seeds which have not yet germinated. (Use cut test.)"

TESTING HOMOGENEITY OF GERMINATION RESULTS

The use of 4 replicates in a germination test enables the tester to measure the degree of variation in the sample. A simple method is to calculate the range of difference in germination per cent between the highest and the lowest of the sub-samples. The range can then be compared with the table published by ISTA, reproduced here as Table (it is applicable to replicates of equal numbers of seeds, not of equal weights).

Provided that the actual range is less than the maximum in the table, the sample can be regarded as homogeneous and the average of the four replicates is accepted. If the actual range exceeds the maximum in the table, then a new sample must be drawn and tested. Common causes of such situations are non-homogeneity of seed and, in nursery tests, fungal or insect damage in one or more of the replicates. Retesting is also necessary when an extremely high percentage of full, ungerminated seeds are left at the end of the test (Bonner 1974). In this case retesting should be done with a different pretreatment to try to overcome dormancy and improve total germination. This table indicates the maximum range (*i.e.* difference between highest and lowest) in germination percentage tolerable between replicates, allowing for random sampling variation only at 0.025 probability. To find the maximum tolerated range in any case calculate the average percentage, to the nearest whole number, of the four replicates: if necessary, form 100-seed replicates by combining the sub-replicates of 50 or 25 seeds which were closest together in the germinator. Locate the average in column 1 or 2 of the table and read off the maximum tolerated range opposite in column 3.

Average percentage germination		**Maximum range**	**Average percentage germination**		**Maximu m range**
1	2	3	1	2	3
99	2	5	87 to 88	13 to 14	13
98	3	6	84 to 86	15 to 17	14
97	4	7	81 to 83	18 to 20	15
96	5	8	78 to 80	21 to 23	16
95	6	9	73 to 77	24 to 28	17
93 to 94	7 to 8	10	67 to 72	29 to 34	18
91 to 92	9 to 10	11	56 to 66	35 to 45	19
89 to 90	11 to 12	12	51 to 55	46 to 50	20

COMBINING PURITY AND GERMINATION TESTS

Purity tests of most commercial seed lots of Eucalyptus are not made because it is difficult or impossible to separate the seed from the chaff of some species. Species in which seed and chaff are extremely similar in size, weight and colour include E. cloeziana, E. regnans and E. delegatensis. other small-seeded genera in which separation of pure seeds is difficult are Alnus, Betula, Populus and Salix. Even when separation is possible in these species, it is very time-consuming. In the seed laboratory in Canberra it takes only 6–7 minutes to set up a test of eucalypt seed with 4 weighed replicates, but to carry out a purity separation and set up a 4 × 100 seed test takes from 20 to 50 minutes, depending on the degree of difficulty in separating out the chaff (Turnbull 1983). The small size of seed also precludes a cutting test to determine the number of full but ungerminated seeds at the end of a test. For these reasons tests are best made on replicates by weight and results recorded as numbers of germinated seeds per unit weight of the impure mixture of seed and inert material. A squash test may also be used to give a rough estimate of viability.

Where seed is sown broadcast, the practising forester is primarily interested in the number of plants he can expect to get from a given weight of the seed lot he receives. Provided he is told that 1 kg of "impure" seed should produce 36,000 germinated seedlings, he is not worried whether this is caused by a combination of 90 per cent purity × 80 per cent pure seed germination or

80 per cent purity × 90 per cent pure seed germination. For local tests, therefore, if seed laboratory staff is insuffient, there is no objection to omitting the purity test even on species for which it would be practicable. In the case of direct sowing into individual containers, however, there are obvious advantages in having separate figures for purity and germination, since sowing is by numbers of (one to several) seeds per container rather than weight of seed per m^2 of bed or tray.

RESULTS AND DISCUSSION OF GERMINATION EXPERIMENT

Germination percentage was high; K values ranged from 76 to 100 per cent during two experimental runs for the four species investigated (89 and 92 per cent for untreated and primed seed, respectively). Germination of perennial ryegrass primed seed (87 per cent in Runs 1 and 2) increased by an average of 12 per cent relative to untreated seed (76 per cent in Run 1 and 80 per cent in Run 2). This species, with the smallest seed, was the only cover crop showing a significant effect of priming on final germination percentage ($P < 0.05$). Final germination percentage in all other species was =87 per cent and not influenced by priming treatment. Seed priming effect on germination in four species of cover crops, cereal rye, hairy vetch, perennial ryegrass, and oriental mustard. The ANOVA is reported for priming treatment effect on each species. Priming impact on the final percentage of seed germination has been shown to be variable and rarely beneficial in earlier studies with wheat and corn. Genotype effects can be pronounced, as shown in earlier studies of wheat where a cultivar with strong emergence characteristics was not influenced by priming while a weakly emerging cultivar responded positively (5 per cent increase). A study of five corn cultivars showed highly variable germination responses to priming, and contrary to expectations, final germination percentage was generally low in primed seed. The type of priming media and characteristics of the species primed can both be influential. We used a germination assay temperature (22°C) and priming media (water) that have been used successfully in previous priming studies, to focus our study on priming treatment duration and variation in species response.

Priming increased the rate of germination for every species studied except cereal rye. In cereal rye, germination was rapid: 10 per cent germination was observed after ~13 h for all seed treatments. In a survey of Michigan potato farmers, cereal rye's rapid germination was widely cited as a positive attribute of this species, contributing to its use as a winter cover crop.

The potential benefits of priming cover crop seeds were demonstrated by the improvements observed in germination timing and rate for vetch, perennial ryegrass, and mustard. The t_{50} in primed relative to untreated seed was reduced by an average of 5 h (Run 1) and 10 h (Run 2) in hairy vetch. Similarly, t_{50} was reduced by 11 h (Run 1) or 20 h (Run 2) in perennial ryegrass and by 6 h in oriental mustard, for primed relative to untreated seed. Rapid germination could improve the ability of seedlings to take up moisture and survive, particularly if soil is rapidly drying out. Studies have shown the importance of early root growth in nutrient uptake, seedling establishment, and biomass production

Species and priming Duration	Run 1				Run 2			
	t_{10} †	t_{50} †	Slope†	K†	t_{10}	t_{50}	Slope	K
	h		% h^{-1}	%	h		% h^{-1}	%
C. rye								
0 h	13.0a‡	18.3a	9.2a	90a	na§	na	na	na
2 h	13.7a	18.4a	9.5a	82a	na	na	na	na
4 h	14.3a	18.8a	11.3a	92a	na	na	na	na
6 h	12.9a	17.3a	11.0a	88a	na	na	na	na
8 h	11.1a	16.9a	8.4a	87a	na	na	na	na
12 h	11.6a	17.8a	7.6a	86a	na	na	na	na
24 h	11.0a	18.0a	6.7b	85a	na	na	na	na
36 h	11.0a	17.9a	7.3a	86a	na	na	na	na
H. vetch								
0 h	25.0a	36.5a	3.7b	94a	23.1a	34.0a	4.8cb	95a
2 h	15.3b	40.4a	2.4b	94a	18.7a	33.9a	3.4c	96a
4 h	23.1a	36.4a	3.0b	98a	16.8ab	31.8ab	3.5c	96a
6 h	22.0a	35.6a	2.9b	98a	12.7b	27.1b	3.5c	93a
8 h	20.7a	29.5a	5.1a	93a	11.4b	19.7c	6.2ab	94a
12 h	19.3ab	26.6b	6.5a	94a	15.6ab	22.2bc	7.9a	96a
24 h	18.8ab	25.6b	7.1a	94a	12.7b	22.5bc	5.5b	98a
36 h	19.1ab	26.0b	6.7a	95a	14.2ab	21.3bc	7.1a	92a
P. ryegrass								
0 h	30.1a	45.2a	2.8b	76cb	40.1a	70.0a	1.5b	80c
2 h	28.9a	43.2a	3.3a	86b	41.2a	63.8b	2.0a	83bc
4 h	28.2a	43.1a	3.2ab	86b	40.3a	62.8bc	2.1a	86b
6 h	28.5a	42.2ab	3.5a	88ab	40.7a	62.8bc	2.1a	84bc
8 h	27.4ab	37.7b	4.4a	83b	37.2b	65.5b	1.7ab	87b
12 h	25.1b	39.9ab	3.2ab	87ab	39.1ab	60.6d	2.1a	85bc
24 h	25.5b	41.7ab	3.0ab	90a	35.8c	61.2cd	2.0a	94a
36 h	25.3b	42.7a	3.1ab	89a	35.7c	63.9bc	1.7ab	88b
O. mustard								
0 h	na	na	na	na	12.2a	18.6a	8.7 b	100a
2 h	na	na	na	na	11.8a	14.5b	20.5a	98a
4 h	na	na	na	na	9.9ab	13.9b	13.9ab	100a
6 h	na	na	na	na	8.7b	13.2b	12.1ab	99a
8 h	na	na	na	na	8.6b	12.4b	14.6ab	100a
12 h	na	na	na	na	10.5ab	13.8b	17.1a	100a
24 h	na	na	na	na	8.2b	12.0b	14.6ab	100a
36 h	na	na	na	na	4.9c	10.3c	9.9b	97a
ANOVA ($P > F$)								
C. rye	ns¶	ns	ns	ns	na	na	na	na
H. vetch	*	*	*	ns	*	***	**	ns
P. ryegrass	*	*	*	**	*	*	*	*
O. mustard	na	na	na	na	**	**	*	ns

Notes: * Significant at the 0.05 probability level.

** Significant at the 0.01 probability level.

*** Significant at the 0.001 probability level.

† t_{10} and t_{50} are the time to 10 and 50 per cent germination, respectively; slope is the rate of germination, andK is the final per cent germination.

‡ Within column means followed by the same letter are not significantly different;

compare within a species.

§ na, not applicable.

¶ ns, not significant at $P < 0.05$.

Previous research documented a rapid response of primed wheat to germination conditions, where approximately 70 per cent of water-primed seeds had germinated after 24 h compared with 25 per cent of untreated seed. However, a second run reported in the same study showed modest (8 per cent) advancement of germination timing, illustrating the variability encountered in seed response to priming technologies. Variation in germination response has been shown to differ by seed lot for the same cultivar. In our study, results from both runs, where a different seed lot was used for each run, were relatively consistent. The degree of consistency from seed lot to seed lot has practical implications, as investment in cover crop priming will only be profitable if reliable benefits accrue.

Soil Core Experiment

This assay tested the consequences of priming for seedling emergence in soil compacted to 1.8 g cm-3, contrasting with the optimal germination conditions of Experiment 1. Final per cent germination in compacted soil (henceforth referred to as emergence) ranged from 40 to 69 per cent, lower than the 80 to 100 per cent germination observed in the moist paper assay of Experiment 1, and potentially more representative of seed response in a stressed field environment. In the compacted soil assay, perennial ryegrass emergence was enhanced in primed versus untreated seed by 20 and 39 per cent in Runs 1 and 2, respectively. This is a substantial increase relative to the 12 per cent increase due to priming observed in Experiment 1.

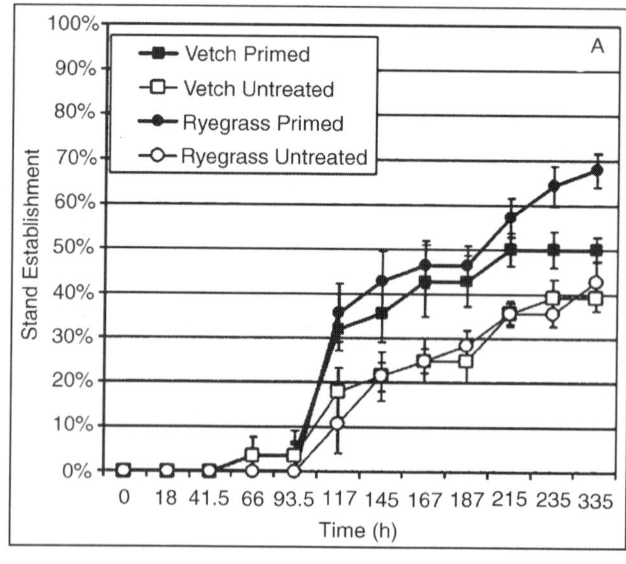

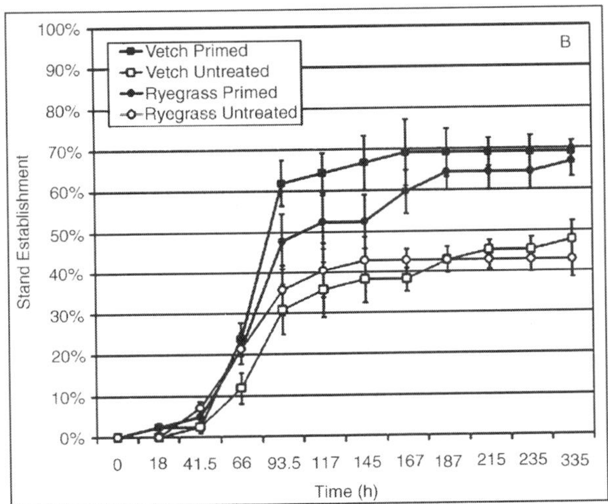

Fig. Seedling Establishment in Cores of Compacted Soil from Seed that was Primed (24 h) or Untreated, for Hairy Vetch (Squares) and Perennial Ryegrass (Circles). Graph 1A is Run 1 and Graph 1B is Run 2; Error Bars Indicate Standard Deviation from the Mean.

Similarly, emergence from compacted soil of primed hairy vetch was improved by 20 per cent in Run 1 and 42 per cent in Run 2, compared with untreated seed. Hairy vetch emergence curves showed increases in t_{50}, slope, and final germination in Experiment 2 ($P < 0.01$, 0.05, and 0.01, respectively). This stands in contrast to Experiment 1, where no priming effect on final germination percentage was observed for hairy vetch.

The results obtained from this compacted soil assay underline the importance of quantifying priming effects on seedling vigour. Other studies have found similar results: canola germination evaluated in Petri dish assays identified minimal priming effects; however, in container-based soil studies, priming increased canola seedling emergence from 31 to 73 per cent for untreated vs. primed seed. Further, in a study of corn response, germination in a moist paper assay was only advanced by 0 to 5 h for primed compared with untreated seed, while priming advanced germination by 20 h or more in a sand culture system.

In our study, we observed that t_{50} was 30 h earlier for primed vs. untreated hairy vetch seed in Run 1; however, more modest results in line with our paper germination assay were obtained in Run 2, where primed seed emerged 9 h earlier than untreated seed. Similarly, in perennial ryegrass, results for compacted soil showed strong responses to priming in Run 1, while Run 2 responses were minimal. The difference observed in the two runs may be attributable in part to differences in seed lot response. Because germination characteristics frequently differ by seed lot, it is important to evaluate different lots.

INDIRECT TESTS OF VIABILITY

Estimating the germination potential of a seed lot by actually germinating a sample of it is often the method most relevant to practical forestry. But the tests take several weeks to complete and for some species pretreatment may take some additional weeks or months. For this reason much research has been conducted to find other methods by which seed viability can be estimated accurately but much more rapidly than by germination testing. The following account of these methods closely follows that of Turnbull.

The objects of quick viability tests are:

- to determine quickly the viability of seeds of species which normally germinate slowly or show dormancy under the normal germination methods;
- to determine the viability of samples which at the end of the germination test reveal a high percentage of fresh ungerminated or hard seeds.

Only two methods, the topographical tetrazolium test and the embryo excision test were previously accepted by the International Seed Testing Association as official methods for some species of seeds. ISTA has recently accepted the X-ray method as a valid alternative to the cutting test for the detection of empty and insect-damaged seeds. The following tests can be applied depending on the circumstances:

CUTTING TEST

The simplest viability testing method is direct eye inspection of seeds which have been cut open with a knife or scalpel. If the endosperm is of normal colour with a well developed embryo, the seed has a good chance of germinating. This test is not very reliable. Seeds with milky, unfirm, mouldy, decayed, shrivelled or rancid-smelling embryos and abortive seeds that have no embryo can be judged as non-viable without much difficulty (Bonner 1974). But it is not possible to distinguish moribund, recently dead or recently injured seeds which still appear the same as sound ones. The cutting test, as already mentioned, is used at the end of a germination test to determine the apparent viability of ungerminated seeds; it is also a useful tool in estimating the size and maturity of the seed crop before collection and the efficiency of methods used in processing.

In the philippines good correlation has been found between cutting and germination tests in fairly large-seeded species such as Leucaena,Intsia bijuga and Lagerstroemia speciosa, but germination per cent was consistently 10–20 per cent less than the percentage of sound seeds on cutting test.

TOPOGRAPHICAL TETRAZOLIUM TEST

The tetrazolium method is only one of a number of biochemical tests which have been developed for seed testing. The various tests have been briefly reviewed by Moore. The tetrazolium test was introduced in 1942 by G. Lakon in Germany. In this method living cells are stained red by the reduction of a colourless tetrazolium salt to form a red formazan. The method emphasizes the need for a knowledge of the soundness of individual embryo parts for predicting the development of embryos into countable seedlings.

The testing procedure is described in detail in the ISTA Rules which approve the test for some species of hardwoods and conifers which germinate slowly by regular germination methods. Normal practice is to soak the seeds in water for about 20 hours, then cut or puncture the seed coat to facilitate entry of the 1 per cent aqueous solution of tetrazolium (TZ) and immerse the seeds in the dark for 48 hours. The process can be greatly speeded up by cutting through the seed at a distance of one third from the micropyle and placing it in a Vitascope vacuum machine for only half an hour. This method gives satisfactory results in Denmark but interpretation of results requires more experience in the operator than the method of seed immersion followed by excision of the stained embryo. The test is done on 4 replicates of 100 seeds each.

Justice states that, while the tetrazolium procedure is good in principle, its practical use in routine testing is limited by many problems, including: difficulty in staining of some seeds; necessity of cutting of dissecting seeds to permit observation of stained parts; poor agreement with results of germination tests in some cases, especially for seed of low germination capacity; lack of uniform interpretation of staining and difficulty in interpreting the significance of different degrees of staining; and an increase in man-hours required to test 400 seeds compared to regular germination tests.

The necessity for an experienced analyst for the successful use of this "common sense test" is admitted by Moore. There is little doubt that the test can be useful for testing the viability of certain species, providing trained staff are available to prepare the seeds and evaluate the results.

EXCISED EMBRYO TEST

By this method, the seeds are soaked for 1 – 4 days and the embryos are then excised from the seeds and placed on moist filter paper or blotter discs in petri dishes. The tests are placed in the light at a constant temperature of 20°C. The condition of the embryos is examined daily. Depending upon the species and lot differences, the tests can be terminated after only a few days, up to a maximum of 14 days, or as soon as distinct differentiation into viable and non-viable embryos can be made.

The excised embryo test is similar to germination tests in that it measures the quality of the seed by their actual germination. In addition it allows some measure of the embryo dormancy to be made, by counting those seeds which, although not growing normally, have grown slightly, remained firm and have kept their colour for the test period. The test is not valid for previously germinated seeds and must not be applied to samples which contain any dry germinated seeds. The success of the test requires considerable skill and experience in the operator and the ISTA rules restrict it to only a few species.

In a comprehensive study, Schubert compared the excised embryo method with the tetrazolium method for determining the viability of dormant tree seeds.

He concluded that the tetrazolium method should receive preference over the embryo excision method but that improvements in the tetrazolium test should be made by providing for the use of bactericides and stronger reducing solutions to resolve doubts in weakly stained tissues.

RADIOGRAPHIC METHODS

Radiography was first used to determine seed quality over 70 years ago. The studies of Simak and Gustafsson highlighted the X-ray technique as a diagnostic method of tree seed analysis. The X-ray contrast method which uses various contrast or radiopaque agents was developed and applied successfully to species of Pinus and Picea.

The X-ray method permits the detection of empty seeds, mechanical damage and abnormally developed internal seed structures, measurement of the thickness of the seedcoat and assessment of the seed viability when combined with a contrast agent.

The X-ray contrast method is based on the principle of semipermeability. When seeds are treated with a contrast agent, for example aqueous $BaCl_2$ or vaporous $CHC1_3$, their living tissues are able to prevent its entry due to their semi-permeability, but the dead tissues become impregnated. The impregnated tissues absorb X-radiation more intensively than the unimpregnated ones and thus appear lighter on the film than the unimpregnated ones. The contrast permits living and dead tissue to be located in the seed and an estimation of its viability. There are now possibilities of using non-toxic water, instead of toxic $BaCl_2$ or $CHC1_3$, as a contrast agent for testing seed viability.

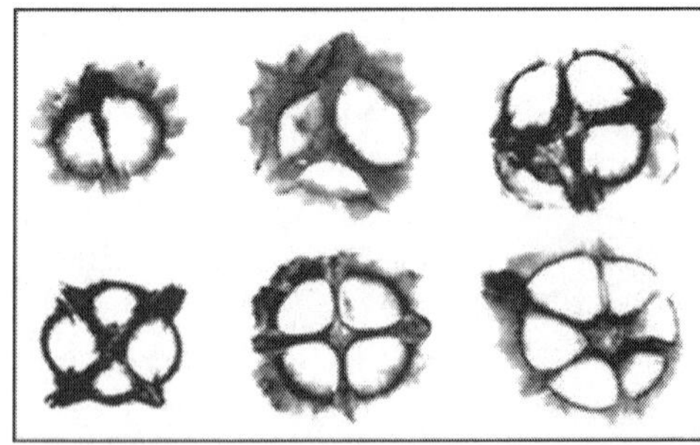

Fig. X-ray Radiograph of teak Fruits showing the Variation in the number of Locules (two to six).

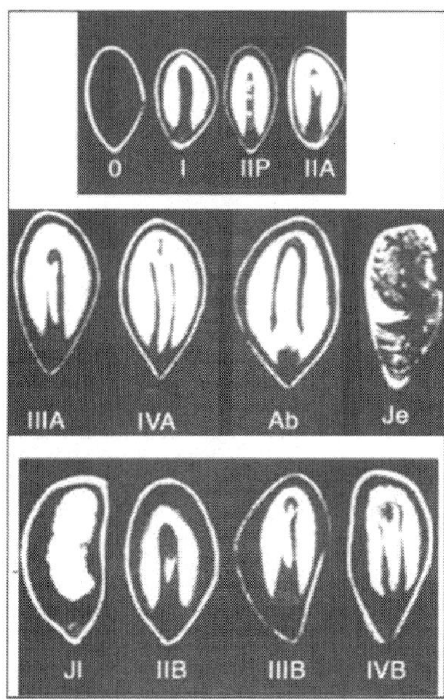

Fig. X-ray Radiographs Showing Embryo and Endosperm classes in Coniferous seeds.

O	Neither embryo nor endosperm (= empty seed)
I	Endosperm, embryo cavity developed but no embryo observed.
IIP	Endosperm, one or more small embryos the length of which does not exceed their breadth ("point embryos").
II	Endosperm, and one or several embryos, none of which is longer than half of the embryo cavity.
III	Endosperm, and one or more embryos, the longest of which measures between half and three quarters of the embryo cavity.
IV	Endosperm, with one fully developed embryo, completely or almost completely occupying the embryo cavity. Diminutive embryos rarely occur.
A	The endosperm almost fills the seed coat to capacity and easily absorbs the x-rays.
B	The endosperm fills the seed coat incompletely and is often shrunken or otherwise deformed. The x-ray absorption is inferior to that of class A.
Ab	Seed with abnormally developed endosperm or embryo.
J	Seeds damaged by insects, containing larvae (JI) or their excrement (Je).

The development of soft X-ray equipment has greatly simplified the operation. Complicated photographic equipment is not necessary and pictures can be made with polaroid film which provides clear and detailed radiographs within 30 seconds. X-ray radiography has been successfully applied in determining the number of seeds in fruits of teak (Tectona grandis) and for studying their degrees of development. The technique has been tried on the fruits or seeds of sixty tropical forestry species and the results show that it can be reliably applied in processing such seeds.

A technique of stereoradiography as a supplement to the X-ray contrast method for use in seed quality testing has been developed by Kamra, Meyer and Wegelius. The chief advantage of stereoradiography is that it is possible for the observer to have a three-dimensional view of the object from a pair of radiographs. In this way, the exact topographical location of the contrast agent in the seed can be reliably determined. This increases the information which can be obtained from radiographs and adds to the analytical accuracy.

The X-ray method is a useful one and is likely to play an increasing role in seed testing. Earlier models of X-ray machines were expensive, but recent models, particularly from Japan are much cheaper and now cost less than a cabinet germinator. Improvements in photographic films and paper have speeded up the process and simplified the interpretation, so that technicians can be trained easily to produce consistent results. ISTA has accepted the method as a valid alternative to the cutting test for the detection of empty and insectdamaged seeds. It also shows considerable promise for distinguishing between viable and non-viable seeds among "full seeds". For certain temperate conifers it has been possible to obtain good correlation between the Development Class (DC) of seeds, based on the development of both embryo and endosperm, and their germinability.

SPECIAL TREATMENTS FOR MECHANICAL DORMANCY

The thick, tough but water-permeable coverings of seeds exhibiting mechanical dormancy prevent embryo growth even when water can be freely imbibed. This mechanical obstruction to germination can be overcome by a period of 'warm moist' treatment the length of which varies according to species.

The treatment recommended by Gordon and Rowe for temperate species is:

- Soak seeds in several times their volume of cold water at approximately 3–5 ° C for 48 hours.
- Drain off the water and mix the seeds with two to four times their volume of a moistened, water-retaining medium such as sand, sand/ peat mixture, vermiculite.
- Store at a warm temperature. A constant 20–25°C or alternating 20° and 30°C is suitable for many species.
- Open containers weekly, mix seeds and, if surfaces show signs of drying out, remoisten with water spray.

The period of 'warm moist' treatment can be shortened in some species by a preliminary treatment in sulfuric acid. This requires more care and expertise than the use of acid for physical dormancy. Seeds or fruits must be thoroughly dried before treatment and the process should be limited to the partial digestion of the outer layers only, leaving the final weakening of the inner layers to be done by the subsequent warm moist treatment. In most cases it is preferable to accept the safer but slower method of warm moist treatment alone. Periods of treatment vary from 2 weeks for some species of Prunus to 16 weeks for some species of Crataegus.

At the end of the appropriate period seeds which possess only mechanical dormancy are ready for sowing. Many species in this class also have physiological dormancy of the embryo and will need further treatment to remove this, as described.

It may be noted that the warm moist treatment which removes mechanical dormancy is identical with the warm moist treatment which removes morphological dormancy (underdeveloped embryos).

TREATMENTS DESIGNED TO BREAK ENDOGENOUS OR EMBRYO DORMANCY

Embryo dormancy is a prominent feature in some temperate genera. For example in the USA Rudolf noted that, of 400 species of woody plants studied, about 60 per cent required pretreatment to induce prompt germination. Endogenous dormancy occurs in both orthodox and recalcitrant seeds. Its occurrence in the lowland tropics is probably rare (most seeds in moist tropical forest germinate quickly or not at all and those in the dry tropics typically have seedcoat dormancy).

It could be of importance in the high altitude tropics and sub-tropics. Trials of prechilling on seed of the Zambales (Philippines) provenance of Pinus merkusii showed no improvement in germination; if anything germination was slightly depressed. Wunder mentions Cordia africana as a species suitable for prechilling for several weeks in moist sand at 5°C but does not give any comparative figures of germination from treated and untreated seeds. In Eucalyptus prechilling is effective in promoting germination in only a few species from the cold temperate region *e.g.* E. delegatensis and E. pauciflora; in the great majority of species it has no effect at all.

Endogenous dormancy includes the cases of embryos which are morphologically underdeveloped at the time of separation from the parent tree and which need a subsequent period for further growth before they can germinate. It also includes the cases of embryos which are morphologically mature at the time of seed dispersal or collection but are physiologically incapable of germination until certain biochemical changes, still little understood, take place.

MORPHOLOGICAL DORMANCY

Seeds with underdeveloped embryos at the time of dispersal or collection will not germinate until the embryos have had time to mature. In a few species maturation is possible during dry storage *e.g.* Gingko biloba. More commonly, a period of moist warm pretreatment is necessary before the embryos develop sufficiently for germination to take place.

The recommended treatment is the same as that for mechanical dormancy *i.e.*

- Soak seeds in several times their volume of cold water at approximately 3–5°C for 48 hours.
- Drain off the water and mix the seeds with two to four times their volume of a moistened, water-retaining medium such as sand, sand/peat mixture or vermiculite.
- Store at a warm temperature. A constant 20–25°C or alternating 20° and 30°C is suitable for many species.
- Open containers weekly, mix seeds and, if surfaces show signs of drying out, remoisten with water spray.

Most species which have underdeveloped embryos also have physiological dormancy, therefore warm moist treatment must be followed by cold moist treatment.

OVERCOMING PHYSIOLOGICAL DORMANCY-COLD STRATIFICATION

Far more common than morphological dormancy among temperate species are the cases where seeds are fully developed at dispersal or collection but are inhibited from immediate germination for physiological reasons. The pretreatment most effective in removing this physiological dormancy is that which approximates to the conditions of overwintering in nature *i.e.* a moist cold treatment, or cold stratification.

Cold stratification will not only overcome physiological dormancy but can also reduce the sensitivity of both dormant and non-dormant seeds to their optimum requirements for light and temperature, resulting in increased rate and uniformity of germination over a variety of conditions. If properly carried out, cold stratification does no damage to non-dormant seeds which are undamaged and have not deteriorated from excessive physiological ageing (Wang in press). It can therefore be safely used when different degrees of dormancy are expected in the same seed lot.

Stratification (sensu stricto) refers to the method of placing seeds in layers alternating with layers of a moisture retaining medium, such as sand, peat or vermiculite, and keeping them at a cool temperature for a period, which is commonly between 20 and 60 days but varies considerably from species to species. The combination of high moisture and low temperature appears to trigger off biochemical changes which transform complex food substances into simpler forms utilized by the embryo when it renews growth at germination. The use of the word 'stratification' has recently been extended to include all forms of cold, moist treatment whether or not the seeds are placed in layers. The present section gives a brief description of stratification in the original sense (outdoor and indoor), while the following section on 'Other moist prechilling methods' describes non-layered methods of cold, moist treatment. The three main requirements for success in both stratification and moist prechilling are a renewable source of moisture for the seeds, low temperature and adequate aeration. Only imbibed seeds will benefit fully from cold, moist treatment, while good aeration is needed to supply oxygen for respiration and to dissipate heat and CO_2. Low temperature not only favours biochemical changes in the seed but reduces micro-organism activity and the risk of overheating and premature germination of the after-ripened seeds.

In temperate climates low enough temperatures to accomplish prechilling are experienced by overwinter stratification in outdoor pits. The method used in the UK is described by Aldhous.

Its main features are:

- The pit should be located in a cool, shady and well-drained site. The bottom 10 cm should be filled with either sand or gravel for drainage. Pit contents should remain moist during stratification, but should never become waterlogged.
- 60–80 cm is convenient for depth and width, the length can be adjusted to the volume of seed to be stratified.
- The pit should be lined at the bottom and sides with a frame carrying 6 mm mesh wire netting as a protection against mice. After the pit is filled, it should be covered by a lid of the same mesh netting.
- The seed to be stratified should be mixed with 4 times its weight of sand and the pit filled with the seed/sand mixture to within 15 cm of the surface. The top 15 cm should be filled with pure sand.
- The start of stratification should be timed in relation to the anticipated sowing date and the optimum period of stratification expected for the species. Seeds should be inspected periodically, starting a few weeks before the sowing date. They should be sown when most seeds are beginning to split and the tips of the radicles are visible but before the radicles have elongated. Delayed sowing leads to broken or injured radicles; hardwoods can sustain such damage without too serious ill effects, but it can cause severe losses in conifers.

Where cold-room facilities are available, stratification may be carried out indoors, where closer control of moisture and temperature can be achieved than with the pit method. A temperature of +1° to +5°C is commonly recommended. In the USA one method in common use is to place 10-to 25-pound (4.5–12 kg) lots of seeds in loosely woven bags which are flattened into disks no more than 3 inches (7.5 cm) thick and alternated with layers of moist medium. Boxes, trays, cans or drums make suitable containers but must have perforated bottoms to facilitate drainage and aeration. Soaking seeds overnight in water at room temperature is common practice with southern pines and hardwoods. Containers should be covered loosely to prevent the seed and medium drying unevenly. The seeds should be inspected periodically to prevent heating, poor aeration and excessive drying and to detect the first stages of germination.

Seeds should be sown soon after removal from stratification. Stratified but ungerminated seeds of some genera, *e.g.* Prunus, may undergo a secondary dormancy if subjected to extreme drying or to temperatures over 20°C. A new cold stratification is then needed to overcome the secondary dormancy.

OTHER MOIST PRECHILLING METHODS

Similar results to stratification in layers may be obtained with many species by storing the seeds moist in polythene bags. As with indoor stratification, seeds should be soaked in several times their volume of water before prechilling, a 48 hour soak at 3–5°C is suitable for many temperate broadleaved species.

After soaking, the water is drained off and the moist seeds are then prechilled at 3–5°C for the period appropriate to each species. Prechilling may be 'naked' *i.e.* without any medium, or the seed may be mixed with 2–4 times its volume of a medium such as moist sand, moist peat or a mixture of the two Polythene bags of about 100 micron thickness make suitable containers since they are moisture proof but somewhat permeable to oxygen. They should be lightly tied and opened weekly when the seed should be mixed and, if necessary, remoistened. A smell of alcohol on opening a bag indicates that anaerobic respiration is taking place because of inadequate oxygen; in this case frequency of opening and mixing should be increased.

Naked prechilling has the advantage that it is easier to check the condition of the seeds throughout the prechilling operation and there is no need to separate seeds from medium at the end of treatment. On the other hand, there is evidence that germination in some species benefits from use of a medium. Gordon and Rowe reported less than 30 per cent germination in 50 days from seeds of Sambucus racemosa treated 'naked', compared with 60 per cent in 20 days from seeds in peat/soil mixture, other elements in the treatments being identical. These authors give detailed prescriptions for pretreatment of a large number of temperature broadleaved species; as a general rule 'naked' prechilling is satisfactory for species which need only a few weeks' prechilling, while use of a medium is advisable for those which need a longer prechilling period and for all species which need warm moist pretreatment.

Periods of prechilling vary considerably from species to species and, to some extent, from seed lot to seed lot within a species. For Abies a period of 3 weeks at 3–5°C has proved satisfactory. The same temperature and period is effective for most of the cold-temperate eucalypts, but some provenances of E. delegatensis need 4–8 weeks for fast, uniform germination. For Nothofagus obliqua and N. procera naked prechilling at 3°–5°C for six weeks, and surface drying before sowing, gave high germination (usually over 80 per cent in 28 days) under nursery conditions. But, as described later in this chapter, treatment with gibberellic acid was a reliable and simpler alternative. In species having deep physiological dormancy the prechilling period may be as long as 20 weeks *e.g.* Liriodendron tulipifera.

For Fagus sylvatica in Poland, after storage for several years at-5°C and 10 per cent MC, the following pretreatments are recommended:

- Allow to defrost.
- Moisten by sprinkling of water and thorough mixing of wetted nuts twice a day for 6 days, at 3°C, until MC rises to 31 per cent.
- Leave wet nuts in unsealed containers without any storage medium at 3°C for a period equal to two weeks longer than the minimum required to produce 10 per cent germination in a sample within two weeks of its transfer to a moist germinating medium. Recurrent sampling and germination testing is needed to estimate this period, which may vary considerably from seed lot to seed lot. MC of 31 per cent to be maintained meanwhile by periodic weighing of containers and remoistening of nuts to restore loss of weight.
- Sow in moist germinating medium at 3°C and leave for two weeks. This should initiate germination of radicles.
- Transfer to temperature of 20°C to promote hypocotyl and epicotyl elongation and seedling emergence, which are inhibited in this species at 3°C.

For large-scale operational sowings, the ideal experimental conditions described above can be simulated to some extent in the nursery by timing the spring sowing so that the nuts experience an adequate period of cold temperature first, followed by the higher temperatures of late spring and early summer.

There is some evidence that steps (2) and (3) of the pretreatment described above can be carried out before, as well as after, storage. The advantage is that the nuts are ready for sowing as soon as they come out of store, without the need for a subsequent cold, moist treatment covering several weeks. Nuts pretreated before storage have been successfully stored for 15 months in France.

CHEMICAL TREATMENT OF PHYSIOLOGICAL DORMANCY

A wide range of chemicals have been tested experimentally in an attempt to overcome internal dormancy. They include gibberellic acid, citric acid, hydrogen peroxide and a number of other compounds. Some have given a degree of improvement, *e.g.* Bachelard found that the germination of dormant seeds of Eucalyptus delegatensis, E. fastigata and E. regnans could be improved by treatment with gibberellic acid (GA). 24 hours' immersion in either GA 3 or GA 4/7 of Nothofagus obliqua has given rapid and complete germination in 14 days, although this normally dormant species otherwise requires 28–42 days' stratification. Shafiq found that the strength of the gibberellic acid had only a small effect, 200 ppm giving 100 per cent germination in 8 days, 50 ppm the same germination in 12 days.

The best stratification treatment (42 days at 3–5°C) yielded 70 per cent in 14 days and 88 per cent in 28 days and the control (24 hours' soaking in distilled water and no prechilling) only 20 per cent in 28 days. The saving in time effected by the GA treatment (1 + 12 days compared with 42 + 28 days) is considerable. Results reported later by Rowe and Gordon showed that GA 4/7 was more reliable than GA 3 as it was less sensitive to temperature during the germination period. Excellent germination was obtained throughout the temperature range 15–30°C, whereas GA 3 required temperatures over 21°C for comparable results.

These successes, however, are the exception. In general, for cheapness and reliability chemical treatments cannot compete with stratification or moist prechilling and they are unlikely to play a major role in normal nursery practice in the foreseeable future.

OTHER TREATMENTS FOR ENDOGENOUS DORMANCY

X-rays, gamma rays, light rays in the red region of the spectrum and high frequency sound waves have all been used experimentally to try to overcome dormancy and stimulate germination. Improvement has been reported in some species including Tectona but it has proved difficult to achieve consistent results and the treatments may induce chromosome damage and other abnormalities concluded that there was far more evidence of unfavourable effects from irradiation of seed than beneficial effects. None of these methods are suitable for practical application at the present time.

TESTING MOISTURE CONTENT

The importance of moisture content in affecting the longevity of seeds in storage has been stressed in Chapter 7. In order to control the operations of drying (or moistening) of seeds in preparation for storage and to check the stability of moisture content during storage, it is clearly essential to have reliable methods of measuring the amount of moisture in a given sample.

Methods of determining moisture content of seeds have been classified by Justice into (a) basic methods in which the moisture is driven out of the seeds by heat and measured by the loss of weight of the original material, or the weight or volume of the condensed moisture, and (b) practical methods designed for rapid routine work and standardized against one or more of the basic methods. Probably all the moisture cannot be driven out of seeds without driving out small amounts of other volatile constituents or causing chemical changes in the material which would result in weight changes. In applying any method, therefore, it is necessary to adhere closely to the prescribed procedure in order that the results of all tests made by that method will be comparable.

Until recently ISTA prescribed three possible procedures:

(1) Drying in an oven for 17 hours at 103°C (2) Drying in an oven for one to four hours at 130°C and (3) Toluene distillation. Method (2) is applicable only to certain agricultural seeds and method (3), previously used for Abies, Cedrus, Fagus, Picea, Pinus and Tsuga, has now been eliminated because it had ceased to be used in practice (ISTA 1981 c). This leaves only method (1) - the "Low constant temperature oven method" as applicable to forest trees.

The test should be made on two samples of about 5 g each, drawn from the working sample including impurities, not on pure seeds. Large seeds should be ground, broken or cut into small fragments to facilitate drying and a good rule of thumb is that any seeds that average over 10 mm in diameter or length should be broken. The samples should be weighed and placed in metal containers, well-spaced to facilitate air circulation, within an oven which is maintained at a temperature of 103° ± 2°C for 17 ± 1 hours. At the end of that period the seed should be placed in a desiccator to cool for 30 – 45 minutes and then reweighed. The relative humidity in the laboratory where the final weighing is done should be less than 70 per cent, to avoid rapid re-absorption of moisture. The difference in MC of the two samples should not exceed a stated tolerance %. If it does, a further pair of samples should be tested; otherwise the mean of the two samples is the final result. The tolerance earlier prescribed by ISTA for all species was 0.2 per cent but, as pointed out by Gordon (1979) and Bonner (1981), a single tolerance figure is not applicable to all species. At the 1983 ISTA Congress in Ottawa, the tolerances approved for moisture tests in tree seeds were agreed as follows:

Sample Condition	**Tolerance %**
Small seeds, moisture <12%, *e.g.* Picea, Alnus	0.3
Large seeds, moisture <12%, *e.g.* Carya	0.4
Small seeds, moisture >12%	0.5
Large seeds, moisture 12 to 25%	0.8
Large Seeds, moisture > 25%, *e.g.* Quercus	2.5

For tropical tree seed laboratories wishing to conform to ISTA rules, this relaxation of tolerances will be a considerable help.

The calculation of moisture content should be made on a wet weight or fresh weight basis *i.e.*

$$\text{Moisture Content\%} = \frac{\text{Original weight} - \text{Oven weight}}{\text{Original weight}} \times 100$$

Although wet weight basis is prescribed by ISTA and is becoming increasingly the standard form for expressing moisture content, it is not yet universal. To avoid any doubt, the method of calculating moisture content should be stated explicitly on any certificate or statement of results. As explained by Gordon and Rowe, provided that the initial fresh weight of a seed lot is measured and the initial moisture content (wet weight basis) calculated by oven-drying a sample, any new MC reached as a result of drying (or wetting) can be calculated directly from the new weight of the seed lot; there is no need for further oven-drying of samples at the new MC. The desired weight of the seed lot to be achieved through drying (or wetting) can be calculated by multiplying its initial weight by the initial dry matter percentage and dividing by the desired dry matter percentage.

---

e.g. (1) If initial wet weight of a seed lot = 50 kg and the MC (wet weight basis), determined by oven-drying a sample, is 25%, the oven-dry weight = 75% of wet weight = 37.5 kg.

(2) If a period of drying reduces the wet weight to 46.5 kg, the new MC

$$\left(\text{wet weight basis}\right) = \frac{(46.5 - 37.5)}{46.5} = \frac{9}{46.5} = 19.4\%$$

(3) If it is desired to reduce the MC (wet weight basis) to 10%, then desired oven-dry weight will be 90 per cent of new wet weight and the seed lot must be further dried until its wet weight

$$\text{must be further dried unit its wet} = \frac{50 \times 75}{90} = 41.67\,\text{kg.}$$

Electric moisture meters give rapid estimates of seed moisture but they are not considered accurate enough for official seed testing. Their rapid operation does make them very useful in certain situations, for example they should be accurate enough to check tree seed moisture as a guide to drying seeds for storage. Meter readings are converted to seed moisture content by means of charts supplied by the manufacturer or developed from calibration curves in the laboratory for the species in question. Most meters will not measure moisture above 15 – 20 per cent and require a minimum of 90 – 100 g of seeds for a test (Bonner 1981). A locally made, cheap and portable electric moisture meter has been used successfully in Thailand for several years to measure the MC of rice grains and could be used for tree seed of similar size. It measures electric capacitance and uses a 9-volt battery as the source of power.

Electric moisture meters are well suited to small seeds, but cannot be used for large seeds such as Juglans or Quercus; winged seeds such as Fraxinus are also difficult to measure. Large or winged seeds can be rapidly dried in a microwave oven. If the oven is preheated, drying can be completed in 5 minutes and weighing in 6 minutes immediately afterwards if in an electronic balance, or after 30 – 45 minutes' cooling in a desiccator if weighing is done in an ordinary balance. Results can be expected to be within 7 per cent at a probability of 0.05 in the case of large seeds of high MC such as Quercus, and within 2 per cent in Fraxinus and Carya, as compared with the more accurate results of the slower conventional methods. A simple and cheap method of drying seeds quickly is to use an infra-red lamp (Gordon and Rowe 1982). A weighed sample is subjected to heat from an infra-red lamp with such an intensity that it loses all its moisture, without being burnt, in about 20 minutes. When the loss of weight ceases, the new weight is measured and the percentage loss calculated.

An up-to-date account of the measurement of tree seed moisture content has been published recently.

SEED EVALUATION

A seed is considered to have germinated after the emergence and development from the seed embryo of those essential structures which are indicative of the seed's capacity to produce a normal seedling under favourable conditions. Abnormal seedlings are not included in the germination count because they rarely survive to produce plants.

The ISTA Rules (ISTA 1976) recognize four groups of abnormal seedlings:

a. Damaged seedlings,
b. Deformed seedlings,
c. Decayed seedlings,
d. Seedlings with unusual hypocotyl development.

These groups and their characteristics are defined in detail in the ISTA Rules. In a laboratory test the majority of the normal seedlings are usually removed at the interim counts, but the assessment of many of the doubtful and abnormal seedlings must be left until the end of the test, to ensure that slower growing but otherwise normal seedlings are not incorrectly classified.

For many species the initial count is carried out one week after starting the test and assessments may be carried out at weekly intervals until the test is ended. More frequent assessment is needed if a more precise picture is needed of the speed of germination. At the end of the test period, all remaining ungerminated seeds should be cut and examined, and the number of fresh, firm and possibly viable seeds recorded (Bonner 1974). At the same time observations may be made on the condition of unsound ungerminated seeds *e.g.* an abnormally high incidence of insect - damaged or mechanically damaged seeds which would indicate the need for improvements in seed hygiene or seed processing methods. The record of the germination test commonly shows the percentage of germinated and the percentage of ungerminated but apparently sound seeds separately, *e.g.*

Germination %	=	82 %
Sound ungerminated seeds	=	6 %
Viability %	=	82 % + 6 % (= 88 %)

It can be seen from Tables that the duration of a test commonly lasts from two to five weeks, but this does not allow any period for pretreatment to overcome dormancy. Many species require no pretreatment and some types of seedcoat dormancy can be successfully treated in a matter of hours. But the prescribed pretreatment of Tectona takes 18 days and some prescribed prechilling treatments 3 – 9 months. When planning a testing programme it is necessary to take account of pretreatment as well as testing time. In extreme cases it may be necessary to replace the germination test by one of the indirect tests of viability described below.

Fig. Acorns of *Quercus Alba* Germinating on Kimpak in USA. Note spacing between seeds.)

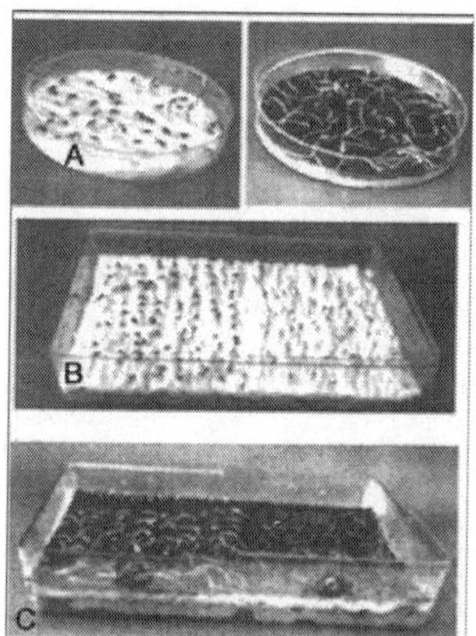

Fig. Germination of Douglas-fir and Lodgepole Pine seeds on Kimpak (A left) and blotter/Kimpak (A right) in Petridishes (A) and Canadian clear Germination Boxes (B-C).

GERMINATION ENERGY

Germination Energy has been defined in more than one way: (1) The per cent, by number, of seeds in a given sample which germinate within a given period (defined as the energy period) *e.g.* in 7 or 14 days, under optimum or stated conditions or (2) The per cent, by number, of seeds in a given sample which germinate up to the time of peak germination, generally taken as the highest number of germinations in a 24 hour period.

In both the above definitions the length of the energy period is considerably shorter than the full test period prescribed by ISTA. Germinative energy is a measure of the speed of germination and hence, it is assumed, of the vigour of the seed and of the seedling which it produces. The interest in germinative energy is based on a theory that only those seeds which germinate rapidly and vigorously under the favourable conditions of the laboratory are likely to be capable of producing vigorous seedlings in field conditions, where weak or delayed germination is often fatal. Little experimental evidence has been published on this theory, but excessively delayed germinants must get eliminated automatically from the nursery, either because they are suppressed by older and more vigorous competitors or because, if transplanting has already been completed, they are not worth the trouble of a special supplementary transplanting. An example of calculating germination energy is given on pp. 234–236.

Another method of comparing germination energy of different seed lots is to record "Rate of Germination" *i.e.* the number of days required to attain 50 per cent of germination capacity. The shorter the period, the greater the germination energy. Yet another method is to evaluate the stage of development of germinated seeds and classify them into a number of development or vigour classes. For example Wang used seven classes for germinated normal seeds of Picea glauca, in addition to ungerminated and abnormally germinated seeds; they ranged from seedlings with healthy root, fully developed hypocotyl and seedcoat completely shed to seeds with seedcoat burst open but as yet no radicle emergence.

GERMINATION VALUE

The concept of Germination Value, as defined by Czabator, aims to combine in a single figure an expression of total germination at the end of the test period with an expression of germination energy or speed of germination. Total germination is expressed as (final) Mean Daily Germination (MDG), calculated as the cumulative percentage of full seed germination at the end of the test, divided by the number of days from sowing to the end of the test. Speed of germination is expressed as Peak Value, which is the maximum mean daily germination (cumulative percentage of full seed germination divided by number of days elapsed since sowing date) reached at any time during the period of the test.Germination Value (GV) can then be calculated from the formula.

$$GV = (\text{final})\ MDG \times PV$$

Germination Value, as an integrated measure of seed quality, has been used by several tropical seed workers *e.g.* for Terminalia ivorensis and for Pinus kesiya.

An alternative method of calculating Germination Value has been proposed by Djavanshir and Pourbeik, who found that it was more closely related to survival of plants in field nurseries than was Czabator's method, in the case of Pinus ponderosa and P. eldarica in Iran. The formula they proposed was:

$$GV = (\Sigma DGS / N) \times \frac{GP}{10}$$

where:

GV	=	Germination value
GP	=	Germination per cent at the end of the test
DGS	=	Daily germination speed, obtained by dividing the cumulative germination per cent by the number of days since sowing
DGS	=	The total obtained by adding every DGS figure obtained from the daily counts
N	=	The number of daily counts, starting from the date of first germination.

The calculations required are somewhat lengthier than those of the Czabator method and, for many patterns of germination, the simpler method is likely to give sufficiently accurate comparisons between seed lots.

5

Cleaning Methods of Seed

The main characteristics by which sound seeds may be distinguished from inert matter including sterile and empty seeds are size and shape, specific gravity, colour and surface texture.

The ease with which sound seeds can be differentiated depends on:

- The degree of difference which exists between the seeds and the matter to be separated from them
- The degree of uniformity among the seeds themselves.

Colour, size and shape are useful criteria for visual separation, while most seed cleaning machines make use of seed size and specific gravity. Screening and sieving methods separate by seed or particle thickness or diameter; the indented centrifugal cylinder by particle length; liquid flotation and blowing, fanning and winnowing methods by specific gravity; while frictional cleaning methods rely on differences in surface texture.

Modern cleaning machines often combine more than one method, so that the cleaning process is both effective and quick. However, the species and the amount of seed to be handled will determine whether cleaning is best carried out by hand, by improvised equipment or by specialist machinery. The following account of cleaning and grading methods is based on Turnbull.

SCREENING OR SIEVING

In most cases a number of sieves with different sized perforations are used and the cleaning is a process of gradually sifting out smaller and smaller particles. It is not only the size of the perforations which determines the quality and quantity of the seed cleaned; other important factors include the precision of the perforations, the angle at which the sieves operate, the amplitude and speed of movement of the sieves, and the correct cleaning and maintenance of the equipment.

Sieves or screens may be made of flat perforated plate or wire mesh, and sometimes they may be three dimensional such as funnelshaped sieves. For small samples hand held sieves are adequate, but in larger scale cleaning a series of shaking screens is commonly used. In Brazil sieving with a mesh size of 32 per inch (approx. 12.5 per cm) was effective in separating seeds from chaff in Eucalyptus grandis. 84 per cent of good seeds were retained and 89 per cent of chaff was eliminated.

SORTING ACCORDING TO LENGTH

Cleaning with sieves relies on the separation of seeds, with diameter the critical factor. Fractionating according to length cannot be done with sieves, but this is possible with an indented cylinder.

In addition to its use for separating good seed from the impurities, the equipment is used in agriculture for separating seed mixtures and can also be used for grading seeds.

The equipment consists of a slightly inclined horizontal rotating cylinder and a movable separating trough. The inside surface has small closely spaced hemispherical indentations. Small material is pressed into the indents by centrifugal force and can be removed. The larger material flows in the centre of the cylinder and is discharged by gravity. Depending on the type of impurities, the seed may be separated via the indentations or by passing down the cylinder.

BLOWING

Cleaning by blowing is a very important and widely used method. It is based on the principle that any object can float in an airstream of sufficient velocity. For separation in an airstream there are three possibilities: falling, floating or rising. The behaviour of the seed and other matter will depend on their weight, their resistance to the flow of air (volume and shape), and the velocity with which the air moves.

The operation of blowing is often referred to as winnowing or fanning. In its simplest form the uncleaned seed is thrown into the air on a windy day. The components separate out and the desired ones are retained. Indoors the airstream from a fan can be utilized.

Hand winnowing has been used successfully in Thailand to separate full seeds from empty seeds of Pinus kesiya (Bryndum 1975). The initial separation was followed by a second winnowing of the discarded fraction. Cutting tests showed that the original full seed percentage of 82 per cent was raised to 98 per cent in the improved fraction, while the discarded fraction (approximately 10 per cent of the total volume) contained only 18 per cent full seed or about 2 per cent of the total full seed in the seed lot. One operator took 8 minutes to winnow 1 kg. Loss of the very smallest and lightest full seed in the separation process is usually not serious since these seeds are likely to germinate slowly and to be low in vigour.

Simple winnowing machines can be easily constructed and Yim illustrates a machine made entirely of wooden components which is used in Korea. Laboratory blowers can be classified as either the pneumatic type—a fan is situated near the air intake and pushes air through the system, or the aspirator type—a fan is situated near the air outlet and pulls air through the system, creating a partial vacuum. Small laboratory cleaners such as the 'Brabant' cleaner, and the 'Kamas' laboratory aspirator are available.

Another seed blower in common use is the South Dakota Blower. The principle of blowing is that a sample of seed, when suspended in an upward stream of air of a certain velocity, will split into a light fraction and a heavy fraction, the light fraction being carried upward and the heavy fraction falling down.

The two fractions are caught apart from each other. The heterogeneity of the heavy fraction can be further reduced by subjecting it to a second blowing at a higher wind velocity. In that way a light fraction, a medium fraction and a heavy fraction are obtained. The blowing apparatus in the South Dakota Blower consists essentially of a centrifugal blower, the outlet of which is connected to the bottom end of a vertical tube of a few centimeters' internal diameter and about 50 cm in length. The sample is held on a fine wire gauze at the bottom of the tube. A built-in valve allows the wind velocity to be set at a rate that has been found optimal for the species. The lighter particles are blown up and trapped by baffles near the top of the tube, while the heavier are retained at the bottom.

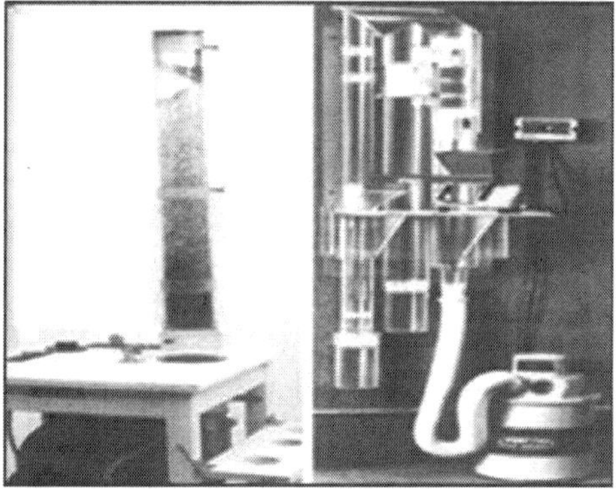

Fig. Electrically Operated Laboratory Seed Blowers: (A) South Dakota Blower (Division of Forest Research, CSIRO Canberra, Photography by Allan G. Edward). (B) Barnes Tree Seed Separator

Fig. Locally Made Seed Cleaner used in Zimbabwe (A) Internal View of Conical Shield Showing Diffusing Baffles (B) In Operatio , Showing the Compartmented Receptacle.

A more complex blower developed in Canada is described by Edwards. It consists of four plexiglas tubes of varying diameters which cause differential air velocity and, by choosing the appropriate combination of tubes, it is possible to use the equipment either to separate seed from chaff or full seeds from empty seeds. It works well with large seeds *e.g.* Abies amabilis but is not suitable for very small, light seeds such as Betula or Chamaecyparis.

In Zimbabwe a home-made cleaner was constructed by fitting a tapered aluminium sleeve with internal baffles onto a constant speed domestic fan (Seward 1980). The narrow end of the sleeve projects over a compartmented receptacle; full seeds fall into the nearest compartment, while the lighter impurities and empty seeds are blown into the further compartments. A similar locally made seed winnowing chamber, which uses a stream of air from an electric fan, has been successfully used in the Solomon Islands to separate debris and empty seeds from dry seed batches of Swietenia macrophylla and Campnosperma brevipetiolata.

Many seed cleaning machines use a combination of winnowing and screening. A coarse upper screen removes larger material, a lower fine screen stops the seeds and lets through fine matter, and then the seed fraction passes through a transverse or nearly vertical airstream from a fan to remove chaff and empty seeds. The air-screen cleaner is the basic equipment of seed cleaning plants. The size of air-screen cleaners varies from the small two-screen model to a modern precision model, which uses several top and bottom screens and in one operation as many as three air separations.

LIQUID FLOTATION

Cleaning by flotation relies on the principle that the density of the seed of a given species is specific both for filled and unfilled seed.

There are two basic methods used:

1. Density method in which liquids with a density or specific gravity between that of the full and empty seed are used. The specific gravity of the liquids used is usually below 1.0 and such that the full seed sinks and the empty seed and light debris float.
2. Absorption method in which water is used and, although both full

and empty seeds float initially, after some time the full seeds absorb water, become heavier and sink. The time of soaking can vary from a few minutes to several hours. This method is useful where there is a very small difference between the specific gravities of the full and empty seeds. The seed must be re-dried after being separated.

The flotation methods can separate out insect attacked, mechanically damaged, and immature seeds from filled mature seeds. The density method can only be applied if a liquid of suitable density is available which is not harmful to the seed. The application and the problems encountered are discussed by Simak (1973); both this and a method for separating viable from non-viable full seeds which was developed by the same author, involving an element of pregermination, are described.

FRICTION CLEANING

Most debris can be removed from the seed by air-screen combinations, but leaf fragments, resin particles and other objects of similar size and density to the seed are difficult to remove.

Friction cleaning relies on the principle that any object falling or sliding over a surface encounters a certain friction. The movement of the particle is proportional to its weight and to a coefficient of friction which depends on the nature of the particle's surface and the surface on which it moves. Separation of seed from debris is made on an inclined cloth or rubber belt on the basis that the angle necessary for the run-off of the seed differs from the angle necessary for run-off of the debris. A continuous upward moving belt removes seeds downwards by gravity and the lighter debris upwards by friction.

SPECIFIC GRAVITY SEPARATION

This method makes use of a combination of weight and surface characteristics of the particles to be separated. It is a method which is finding increasing use in separating and grading tree seeds.

The specific gravity (SG) separator employs a flotation principle. A mixture of seeds is fed onto the lower end of a sloping perforated table. Air, forced up through the porous deck surface and the bed of seeds by a fan, stratifies the seeds in layers according to density, with the lightest seeds and particles of inert matter at the top and the heaviest at the bottom. An oscillating movement of the table causes the seeds to move at different rates across the deck; the lightest seeds float down under gravity and are discharged at the lower end, whilst the heaviest ones are kicked up the slope by contact with the oscillating deck and are discharged at the upper end.

Deck coverings of linen, plastic, and wire mesh have been used to distribute the air uniformly beneath the seeds and provide the correct push to the heavier seed in contact with the deck. A closely woven covering such as linen gives the best results for small seeds. On modern gravity separators, it is possible to control the feed speed, the tilt of the table in two directions, the speed of oscillation and the force of air differentially at various points along the deck. The combination of these different controls makes it possible to adapt the machine to handle a wide variety of species and seed lots.

The specific gravity separator will separate particles of the same density but of different size, and particles of the same size but of different densities. It will not separate efficiently particles which differ both in density and size, *i.e.* separate a larger but less dense particle from a smaller but denser particle. It has been found practical for cleaning the chaff from some eucalypt seeds and for grading pine seeds. Purity% of uncleaned seed of E. grandis after extraction is about 10 per cent. This was raised to 95 per cent purity with 95 per cent germination by SG separator treatment.

DRYING OF ORTHODOX SEEDS

For medium- or long-term storage of many species, a moisture content of 4–8 per cent is recommended. This is considerably less than the MC of freshly collected seeds. Reduction of MC can be achieved in most species by placing the seeds in an ambient atmosphere of relative humidity (RH) 15–20 per cent for a period sufficiently long to allow the seeds to reach an MC in equilibrium with the RH.

The effectiveness of air-drying depends on local climatic conditions. Reduction to an MC of 12–18 per cent is frequently possible, provided that attention is paid to adequate aeration of the seed. Reduction to less than 8 per cent is impossible in most temperate situations and in some areas of the wet tropics, because average RH remains too high. Thus RH in moist tropical West Africa is commonly over 80 per cent in the wet season and over 70 per cent in the "dry" season. An MC of less than 8 per cent is unlikely to be achieved in these conditions. In areas where insolation is high, many species can be successfully dried to 6–8 per cent MC by exposing them to direct sunlight, because the seeds and surrounding microclimate heat up and RH is consequently reduced. Also, for a constant RH, the equilibrium MC of seeds decreases with increasing temperature. Care must be taken to ensure that the seeds are as dry as possible prior to exposure and that they are moved frequently. In Honduras this method works well with Pinus spp. but is not recommended for Cordia because seed of this genus dries too quickly and can reach 4 per cent MC, with damage to the tissues.

As pointed out by Harrington, artificial drying can be accomplished in two ways. One method is to raise the temperature of the air which, provided that no additional water vapour is introduced from outside the system, automatically reduces the RH. The other is to remove moisture from the air without changing the temperature, which also reduces RH. He quotes the example of air at 5°C and 90 per cent RH which is heated to 35°C. Its RH is thus reduced to 15 per cent and it can be blown through the seed by forced ventilation until the latter reaches equilibrium MC. On the other hand, air at

30°C and 90 per cent RH, typical of moist tropical areas in the rainy season, would still have an RH of 40 per cent even if heated to 45°C.

High temperatures can be extremely injurious to seeds, especially those which have a high MC. Generally, drying temperatures should not exceed 40°C and recent trends have been towards lower temperatures and greater air-flows to ensure safety in drying. Barner and CATIE recommend a temperature not exceeding 30°C in the early stages. One possibility is to dry seed in two stages, the first to about 11 per cent MC using a temperature below 40°C, the second to about 5 per cent using a temperature of 60°C.

Provided that the MC is reduced to about 11 per cent in the first stage, a second stage temperature of 60°C is considered safe for most agricultural species. On the other hand, there is some evidence that a high drying temperature, which does not affect immediate germination, may still affect subsequent longevity. For long-term storage for genetic resources purposes, a combination of low RH and low temperature (15 per cent RH and 15° C) is recommended.

Where climatic conditions make it impossible to achieve a low enough RH by heating the air, it is necessary to reduce it by removing water vapour without raising the temperature.

This can be done by either:

- Refrigerating the air to below the dew point, condensing the water vapour on the cooling coils and then reheating the air to 35°C or
- Blowing air through a chemical desiccant, which removes the water vapour, and then through the seed.

Various desiccants are available *e.g.* silica gel, CaO, H_2So_4, Lithium chloride or anhydrous $CaCl_2$ but the most indestructible and easiest to reuse is silica gel. A good example of one drying technique is that furnished by the seed bank of the Regional Genetic Resources Project at Turrialba, Costa Rica. Because of the permanently high air humidity and high day temperatures there, hot air drying is not possible without damage to the seeds. Therefore a silica-gel type dryer, which maintains an RH of less than 15 per cent at 25°C, is used, situated outside the drying room and ducted into it.

Small lots of seed that have first been air-dried to an MC below 20 per cent may be placed in a sealed container with an equal quantity of silica gel that has been freshly dried at 175°C and cooled. The silica gel, seed and enclosed air will come to an equilibrium suitable for storage. See also the section on pp. 156–157 on "Use of dessicants in containers". The period of time needed for forced-draught air circulation to dry seeds to equilibrium moisture content is partly dependent on the accessibility of individual seeds to the air current. Seeds should be spread in thin layers on trays, with space for air circulation between the trays.

The MC of coniferous seeds, which have been extracted from the cones during kiln-drying, may already be close to that recommended for storage. But moisture may be reabsorbed during cleaning and dewinging, and in some species it may be added deliberately to reduce damage during these operations. Therefore MC must be tested and, if necessary, further reduced immediately before storage. At the same time, if cleaning and dewinging are carried out in warm, well-ventilated rooms, this will decrease the amount of drying needed subsequently.

MIXING BEFORE STORAGE

When a large seed lot is to be stored in several containers, it is desirable that seed homogeneity between containers be maintained as far as possible, so that each container is equally representative of the seed lot as a whole. If seeds have been size-graded, mixing is carried out on the ultimate seed lots separated by the grading operation.

Thus, a given seed collection may be graded into a "large seed" fraction and a "small seed" fraction. Each fraction becomes a separate seed lot with its own seed lot number and separate mixing within each of these two seed lots may be done before storage to ensure homogeneity between containers.

GRADING METHOD OF SEED

Within a single species there is a variation of seed dimensions due to environmental influences during the development of the seed and to normal genetic variability. The performance of the seed immediately after germination is related to seed size and, in order to produce a crop of seedlings which will emerge and grow evenly in the nursery, size grading of seeds can be a useful practice.

Size grading can also assist mechanical sowing of seeds. The practice should be used with caution in the case of seeds collected from seed orchards having a restricted number of clones. Since part of the variation in seed size and shape is genetic, grading of seed orchard seed could lead to excessive genetic differentiation and loss of genetic diversity within each of the fractions obtained by grading.

Fig. Air/Screen Seed Cleaner used in Humlebaek, Denmark

Fig. Damas Gravity Seed Separator

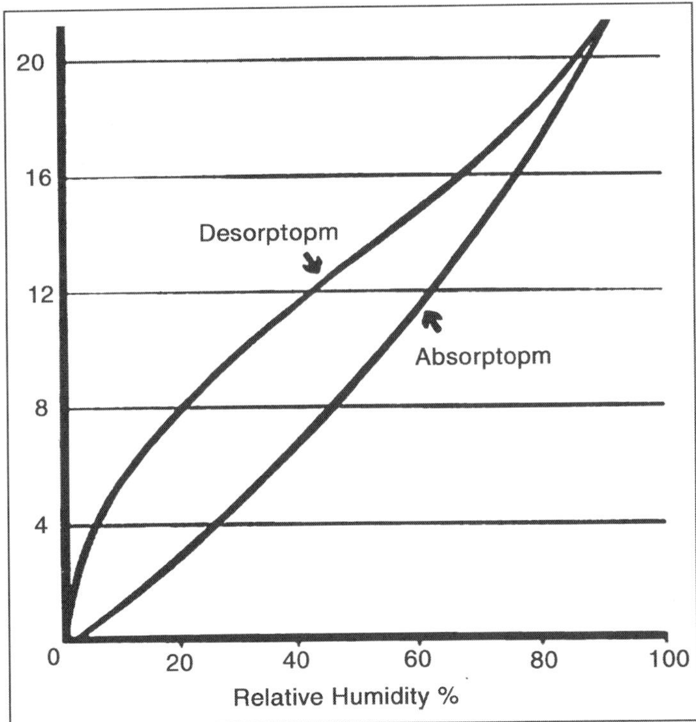

Fig. Equilibrium Moisture Content of Wheat Seed, Showing Separate Curves for Desorption and Absorption.

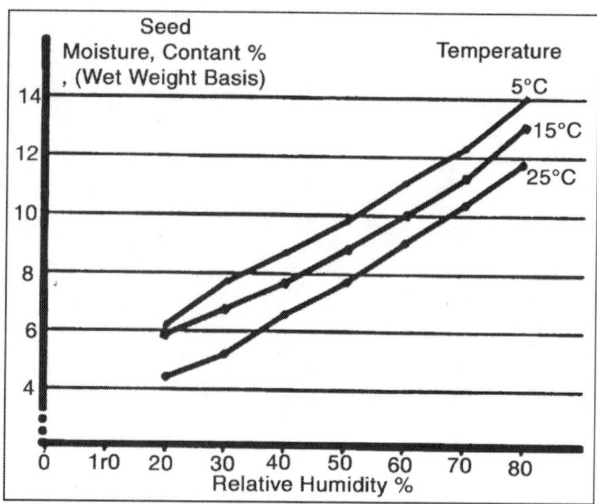

Fig. Moisture Content Percentages of Fresh Seed of *Pinus Palustris* in Equilibrium with Air at Various Temperatures and Relative Humidities

The methods of grading seed vary little from those used in the cleaning process. Sieving and screening, cylinders, air blowing, flotation and specific gravity separation can all be used effectively for size grading tree seeds. Although grading itself is a relatively simple operation, seeds of certain species have to be properly processed before they can be graded, for example the spongy exocarp of teak fruits and the calyx tubes of dipterocarps must be removed in order to get the full benefits of grading. The cleaned stones of Gmelina arborea have been graded by using square-mesh screens of 7,9 and 11 mm mesh; germination varied from 84 per cent for the smallest to 111 per cent for the largest size class (there are usually 1 to 3 seeds per stone).

CONTROL OF MOISTURE CONTENT

After seeds have been cleaned and graded, they are ready for sowing in the nursery. If, however, they are to be put into storage, it is necessary to check their moisture content (MC) and, if necessary, adjust it to the optimum level for storage of the species in question. Adequate facilities for testing MC should be available in the seed processing depot.

For orthodox seeds, which comprise most coniferous and many hardwood seeds, adjustment of MC if needed means further drying. This is described in the following section. Much less commonly and only in the case of recalcitrant seeds which must be stored at a high MC it may be necessary to moisten the seeds in order to raise MC to the optimum for storage.

For example promising results have been achieved for Acer pseudoplatanus by soaking the seeds in water for two or three days and immediately afterwards freezing and storing them at about -7°C in plastic sacks. Other genera, such as Quercus and Castanea, which have been lightly dried under cover in order to loosen their husks or involucral attachments, may benefit from soaking to restore MC to the optimum level (*e.g.* 40 – 45 per cent for Quercus robur, Holmes and Buszewicz 1956, Suszka and Tylkowski 1980), before being placed in moist, cool storage.

RELATIONSHIP OF SEED MOISTURE CONTENT TO ATMOSPHERIC HUMIDITY

Seeds, like cones and fruits, are hygroscopic materials and, when detached from the parent tree, they lose or gain moisture to or from the surrounding atmosphere until their moisture content (MC) reaches a point of equilibrium with the humidity and temperature of the surrounding air. This is known as the equilibrium moisture content (EMC). Once it has been reached, it will be maintained as long as the humidity and temperature of the air remain constant; if they change, the seeds will again lose or gain moisture until a new EMC is reached. Wood is another good example of a hygroscopic material and behaves in a similar way.

"Wet" seed surrounded by "dry" air will lose moisture and therefore weight, while "dry" seed surrounded by "moist" air will gain them. In order to devise the most suitable methods of drying and storing seed, it is necessary to be able to quantify the moistness of both air and seed.

AIR HUMIDITY

Moisture in the atmosphere is in the form of water vapour, but air can hold only a limited quantity of water vapour. If this is exceeded, the air is said to be saturated and the excess moisture condenses as dew. The exact weight of water vapour (WV) which the air can hold at saturation depends on the temperature, as shown in the following table:

Temp. °C	-10	0	10	20	30	40	50	60
Weight of WV at Saturation (g WV Per Kg of Dry Air)	1.6	3.8	7.6	15	27	49	87	152
Density of Dry Air (Kg Per m^3) at a Pressureof 760 mm	1.34	1.29	1.25	1.20	1.16	1.13	1.09	1.06
Weight of WV at Saturation (g WV Per m^3 of Dry Air)	2.1	4.9	9.5	18	31	55	95	161

Most of the time the content of water vapour in the air is less than that at saturation. Relative humidity (RH) is defined as the ratio (usually expressed as a percentage) of the quantity of water vapour actually present in the atmosphere to the quantity which would saturate it at the same temperature; alternatively, as the actual vapour pressure in the air as a percentage of the saturation vapour pressure at the same temperature. For the seedsman relative humidity is the most important measure of atmospheric humidity because the equilibrium moisture content of seeds is most closely correlated with it. For example the MC of seeds will be very nearly the same when in equilibrium with air at an RH of 50 per cent, whether the air temperature is at 10°C (absolute humidity or weight of water vapour present = 7.6/2 = 3.8 g/kg dry air) or at 50°C (absolute humidity or weight of water vapour present = 87/2 = 43.5 g/kg dry air). Though the absolute humidity of one is over ten times that of the other, the relative humidities are the same and it is relative humidity which has the greatest effect on the EMC of seeds. The importance of RH to the EMC of seeds and the dramatic effect of temperature changes on RH explain the importance of heat in the drying of many seeds. From the table above it can be seen that air with 3.8 g water vapour/kg at a temperature of 0°C would be saturated; at 100 per cent RH it would be useless as a medium for drying seeds. But if the same air were heated to 30°C and provided no additional moisture were introduced from outside the system, its RH would be reduced to 14 per cent and it would become a highly effective drying medium.

MOISTURE CONTENT OF SEEDS

The amount of moisture in seeds is usually expressed as a percentage of their weight. Methods of measuring MC are described.

Moisture content can be expressed in two ways:

1. The weight of water expressed as a percentage of the initial "wet-weight" or "fresh-weight" of the seeds (= dry matter + water) or
2. The weight of water expressed as a percentage of the final oven-dry weight of the seeds (= dry matter only).

One of the greatest difficulties in the understanding and application of published results on moisture content derives from the fact that, in the past, both the "wet-weight" and "dry-weight" methods have been used, often with no indication as to which of them were applied in a particular case. According to the ISTA Regulations seed moisture content should always be expressed on a wet-weight basis. For orientation both formulae are given here with a conversion table.

Because of the limited amount of water vapour which it takes to saturate air, a relatively small quantity of seed can hold as much moisture as a great deal of air. One litre of seed dried from 50 per cent to 9 per cent MC (wet weight) at 30°C would lose about 450 gms of moisture to the surrounding atmosphere, enough to change the RH of about 15 m^3 (or 15,000 times its own volume) of air from 0 to 100 per cent. In the case of sun-drying, the atmosphere is so vast that it can absorb this moisture without difficulty, but in an enclosed building the ambient air can quickly become saturated.

$$\text{Percentage of Moist Content, Dry Weight Basis} = \frac{\text{Weight of Water}}{\text{Weight of Dry Matter}} \times 100$$

$$\text{Percentage of Moist Content, Dry Weight Basis} = \frac{\text{Weight of Water}}{\text{Weight of Water} + \text{Weight of Dry Matter}} \times 100$$

Moisture Content,% of Dry Matter (Dry Basis)

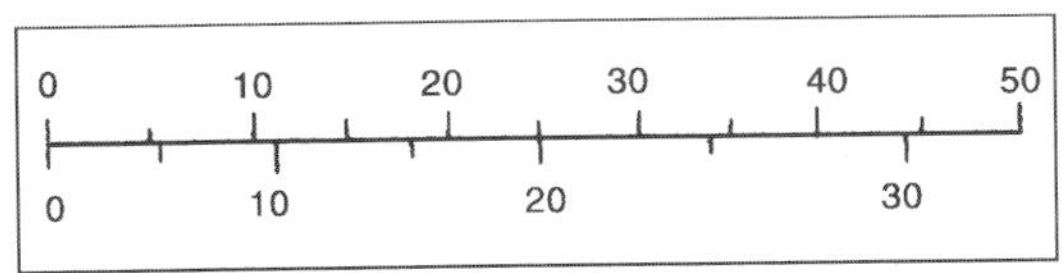

Fig. Moisture Content,% of Total Weight (Wet Basis)

This explains why so much emphasis is put on adequate ventilation in kiln drying, in order to ensure that moist air is removed as it approaches saturation and is replaced by fresh, dry air. The same feature is an advantage when storing dry seeds in sealed containers. Provided that the seeds are correctly dried and the containers properly sealed, a relatively small volume of seed will come into equilibrium with a much larger volume of enclosed moist air without increasing its own MC significantly. If one litre of seed of oven-dry specific gravity 0.5, dried to 9 per cent MC (wet weight), were to be enclosed in a 10 litre sealed container with 9 litres of moist air at 100 per cent RH at 20°C, the total moisture content of the air would be only: $\frac{9 \times 18}{1000} = 0.16\text{g}$ Even if the seeds were to absorb all this moisture, it would only raise their MC from 50 to 50.16 g or MC per cent from 9.09 to 9.12 per cent. The common prescription to fill sealed containers as full as possible with seed is a sound one, but it is based on the deleterious effects on many species of the oxygen in the enclosed atmosphere, not of the water vapour.

OTHER FACTORS AFFECTING EMC

Although relative humidity is the most important single factor affecting the equilibrium moisture content of seeds, it is not the only one.

Temperature

As already explained, temperature has a large indirect effect on EMC because, if absolute humidity is kept constant, relative humidity is directly related to temperature.

It has an additional effect because EMC varies slightly with temperature even when relative humidity remains constant. The effect varies with species but very few data on forest trees have been published. An example quoted for an agricultural crop, sorghum, by Justice and Bass shows that at 50 per cent RH the EMC varies from 12 per cent at 49°C to 14 per cent at -1°C. The difference is slightly greater in some other crops, but in all cases EMC decreases with increasing temperature and constant RH (even though absolute air humidity increases with temperature at the same RH).

Absorption and Desorption

For any species there is a difference of 1 – 2 per cent in the EMC according to whether a moist seed is losing moisture to a drier atmosphere (desorption) or a dry seed is gaining moisture from a moister atmosphere (absorption). The EMC is always higher on desorption and it is the desorption curve which is important in the common situation of drying orthodox seeds from a higher to a lower MC for storage.

Variation in EMC According to Species

The MC of seeds in equilibrium with a given RH and temperature varies with species. The EMC for each species must be determined by trial. An important component in interspecific variation is the percentage of oil content in the seeds. Seeds which store most of their food reserves as proteins or starch have a higher EMC at a given RH than seeds which store food as fats and oils, because the former are relatively hydrophilic, the latter hydrophobic. Among agricultural seeds wheat, with a low oil content of 2 per cent and an EMC at 45 per cent RH and 25°C of 10.4 per cent, may be compared with Brassica oleracea, with a high oil content of 35 per cent and an EMC under the same conditions of 6.0 per cent.

Equilibrium Moisture Contents for 3 Orthodox Species

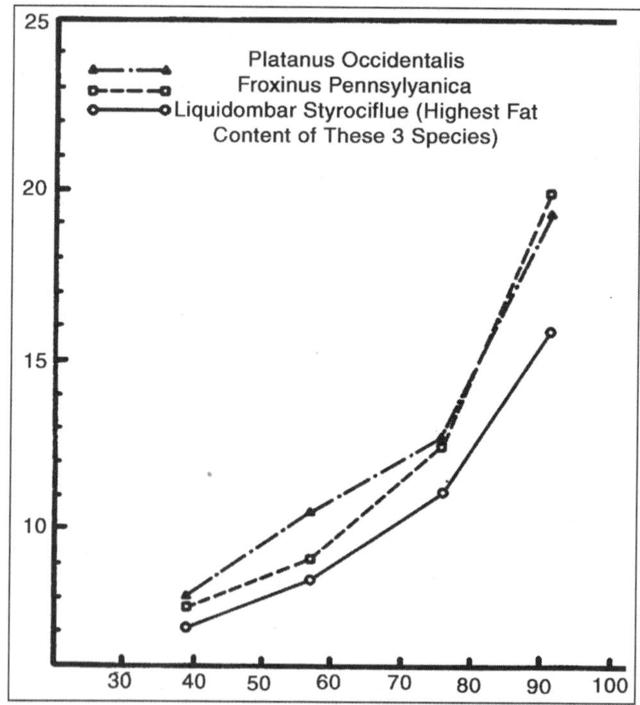

Fig. Equilibrium Moisture Contents for 3 Orthodox Species.

Equilibrium Moisture Contents for 4 Recalcitrant Species

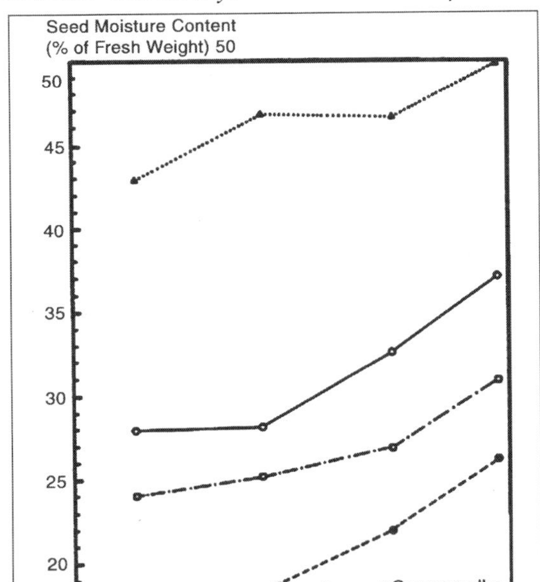

Fig. Equilibrium Moisture Contents for 4 Recalcitrant Species.

Detailed information on the EMC of tree species is sparse, and lacking entirely for tropical species. Some selected examples are given in the following table and in the attached graphs provided by F.T. Bonner.

Table. EMC% (Wet Weight Basis)

Species	Temperature	RH 10	20	30	40	40–55	50	60	70	95
Fraxinus sp.	not given	4.1	6.0	7.4	8.8	–	10.3	12.0	13.9	–
Picea abies	not given	2.4	4.2	5.5	6.7	–	7.8	9.0	10.4	–
Pinus taeda	4 – 5 °C	–	–	–	–	10	–	–	–	17

The graphs show EMC for four recalcitrant (Quercus) species and three orthodox broadleaved species. It should be noted that the EMC among the Quercus spp. is correlated positively with carbohydrate content and negatively with fat content. Q. alba has the highest EMC, the highest carbohydrate content and the lowest fat content, followed by Q. muehlenbergii, Q. shumardii and Q. nigra. Similarly, Liquidambar has the lowest EMC, the lowest carbohydrate content and the highest fat content of the three orthodox species. The same process of tissues reaching an MC in equilibrium with the ambient RH occurs during the drying of fruits for seed extraction as in the drying of the seeds themselves for storage. Usually the exact EMC is less critical in drying fruits, since shrinkage, splitting or scale-opening take place over a range of MC and the process is only carried on to the stage when the fruits release their contained seeds.

OTHER CLEANING METHODS

A number of other methods of seed cleaning have been used experimentally, but are not yet in widespread operational use. They include electronic and electrostatic separators, magnetic separators, electronic colour separators and shaking tables, which separate seeds by the angle at which they rebound when thrown against fixed walls.

Seeds of Ochroma can be successfully cleared of their floss by placing the uncleaned mass on a wire sieve of 0.3 mm mesh and setting fire to the floss. The fire flashes across and the seeds drop through the mesh. Good results are obtained when the seeds are allowed to drop through the mesh into a pan of water. The inflammable oil in the floss burns with intense heat and experience in Honduras has shown that the floss must be spread out thinly to avoid damaging the seed. This method has also been tried to clear Populusseeds of floss but may damage more than 50 per cent of the seeds in this genus.

Seeds of Prosopis are often embedded in a gummy matrix within the pods. *One way of obtaining clean seeds is:*

- To remove one side of the pod mechanically with a knife
- To soak the pod contents in a 0.1 normal solution of hydrochloric acid for 24 hours
- To wash in water for one hour, then dry in the direct sunlight
- To beat or pound the dried mass to separate the clean seeds from the gummy coating. This method has been successful in India, yielding clean seeds with a germination rate of 65% in 12 days.

6

Hybrid Seeds

Hybrid varieties have been evolved in those high valued vegetable crops which exhibit marked heterosis such as solanaceous vegetables (tomato, eggplant, sweet pepper), cucurbits (melons, watermelon, cucumber, squash, pumpkin and gourds), cole crops (cabbage and cauliflower), root and bulb crops (onion, radish, carrot) and fruit vegetable like okra.

SOLANACEOUS CROPS

TOMATO

Manifestation of Heterosis

Hybrid vigour in tomato is manifested in the from of earliness, total yield, uniformity of produce, greater plant vigour and better adaptability to unfavourable environment. In general heterosis for yield in hybrids ranged from 30-50 per cent.

Hybridization Techniques

For commercial hybrid seed production the emasculation of female flowers and pollination by hand is still considered economical and efficient. Most of the seed companies do hybridization this way. Indeterminate tomato varieties are staked and trained with either single stem or double stem, whereas, the determinate tomato varieties are trained with 3 stems.

Usually 1st to 4th cluster on each branch are selected for emasculation. Buds where the corolla leaves have just opened and form an angle up to 45" in respect to the flower axis (a day prior to anthesis) are selected for emasculation. The emasculation operation involves in holding the corolla at the base and with a single upward pull pick off the corolla along with all the stamens.

Usually the anthers are picked off a day before anthesis with the help of forceps leaving the petals intact. Such petals turn yellow on the day of anthesis. Fresh pollen collection on the day of anthesis by a vibrator has the highest viability, because only ripe pollen are shed by vibrating the flowers.

Pollens are collected in a glass tube or on a glass plate from the male line and are transferred to the stigma by finger or by inserting the stigma into a glass tube containing the pollen grains. Left over pollen grains in the glass tube are not used on the next day because its viability is reduced considerably. Half of the calyx of pollinated flowers is removed to distinguish it from unpollinated flowers. Pollen grains can be stored for a longer period (2 months) of time when its moisture content is reduced using a desiccator and the temperature is kept around 0°C.

Physiological Parameters Affecting Tomato Seed Yield

Growth of tomato plant is satisfactory upto 25°C. In tropical regions, prolonged high temperatures (35-42°C) adversely affect pollen fertility and physiology of fertilization, of inflorescence, different altitude and latitude and fertilizer application exert an influence leading to poor seed set. Age of pollen and stigma, abundant or scarce pollination, height on the vigour of the plant.

Improvement in Methods of Hybrid seed Production

Nearly 40 per cent of the total labour expenditure is on flower emasculation during the course of hybrid seed production, which can be reduced by using male sterile lines. Male sterility controlled by recessive genes (totaling 42 in number now) and stamenless mutant controlled by recessive gene (s1) having normal corolla and free stigma, are quite accessible for pollen application. Limited success has also been obtained in phenotypical restoration of sterility in 'ms' forms with silver nitrate treatment. Partial success has been achieved in restoration of (ms) sterility using GA 3 or GA 4/7. Such restoration will result in the production of 100 per cent fertile plants from male sterile forms.

Seed Extraction and Drying

Tomato seeds are extracted mainly by fermentation method. Under warm conditions, the fermentation process is complete in 24 hours. At 25ºC, it requires 2 days for the completion of fermentation process. The pulp is stirred several times in a day to maintain a uniform rate of fermentation and to avoid discolouration of the seed. Seeds are than washed with clean water. Fermentation method of seed extraction also controls bacterial canker disease which is seed born. Tomato seed is also extracted using acid (HCl) or alkali (NaOH), 10 cc or 36 per cent HCI or 30 per cent NaOH is added in 4 kg of tomato pulp. The treatment is given for a period of 15 minutes, which separates the jelly from tomato seeds. The seeds are than washed thoroughly and then dried.

Seed Yield

One kg of tomato fruit will produce 3-4 g of seed yield (1000-1200 seed). Av. seed yield: 60-70 kg/ha depending upon the performance of parental lines.

EGGPLANT AND SWEET PEPPER

MANIFESTATION OF HETEROSIS

In general heterosis in sweet pepper ranged from 35-40 per cent whereas, in eggplant it ranges from 50-150 per cent. Rashid et al. reported 50 per cent heterosis in eggplant in Bangladesh. Heterosis in eggplant is manifested in earliness, fruit number per plant and fruit weight. In case of sweet pepper, heterosis is manifested for plant height, days to flower, fruit weight, number of fruits per plant, early and total yield.

Hybridization Techniques

Emasculation and hand pollination is the useful production techniques. Stigma is receptive a day prior to anthesis in eggplant. Hence bud pollination is possible giving good fruit set and seed yield. In sweet pepper, emasculation is done a day prior to anthesis, whereas, pollination is done in the morning on the day of anthesis. Natural cross pollination ranged from 0.2-46.8 per cent in eggplant flowers. Emasculated flowers are never visited by pollinators.

Pepper flowers are visited by honey bees occasionally. Fresh pollen grains are collected on the day of anthesis by a vibrator and can be stored for a period of 1 to 2 months at *0ºC*, using silica gel for proper drying of the pollen grains. In order to obtain optimum yield and good quality seed, it is essential to train the eggplant as well as sweet pepper plants.

The first and 2nd flowers are harvested at the initial stages. This will boost plant growth as well as the number of seeds formed in subsequent fruits. The training of eggplant involves to allow two lateral branches below the first flower and the remaining lateral branches are removed. This technique is aimed to attain sound growth of plant.

Physiological Parameters Affecting Seed Yield

The eggplant is photoinsensitive but requires an optimum day temperature of 25-30°C and 20-27°C night temperature for its growth. In case of pepper effect of low temperature (8-10°C) on seed set is greater before anthesis than afterwards. Plant growth is satisfactory at day temperature ranging from 20-25°C. Pepper plant prefers high humidity for its proper growth. Providing partial shade prior to flowering boost its vegetative growth. Capsicum flowering and fruit set require bright sunlight.

Use of Male Sterility in Hybrid Seed Production

Male sterile lines are available in eggplant (genic male sterility) and sweet pepper (genic and cms) but are not successful at commercial seed production. Seed set on male sterile lines ranged from 46-67 per cent in capsicum. Use of functional male sterility in hybrid seed production, sweet pepper has a bright future. Cytoplasmic male sterility is being used now in case of chilli pepper to produce F1 hybrid commercially by several seed companies.

Seed Extraction

Eggplant fruits are harvested 50-55 days after anthesis and are stored for a period of 10 days for postharvest repening. Sweet pepper fruits are harvested 60-65 days after anthesis. The ripe fruits are crushed and seeds are separated by washing with excess of water without fermentation. The seeds are dried using dry air at 28-30°C.

Seed Yield

A satisfactory seed yield in eggplant is 150-200 kg/ha with a thousand seed weight ranging from 4-5 g. In peppers the seed yield varies from 100 to 200 kg/ha. One kg of sweet pepper will yield 5-7 g of seed with the thousand seed weight equals to 5 g.

CUCURBITACEOUS CROPS

Sex Expression and Sex Forms in Cucurbits: Cucurbitaceous plants produce three types flowers depending on species and variety. One plant my contain more than one type of flowers, and more than one sex forms may be available in a species. According to the sex forms, cucurbitaceous plants can be grouped as follows:

- *Hermaphrodite*: All flowers of a plant are bisexual (available in some varieties of Luffa, Cucumis and Benincasa spp.)
- *Monoecious:* Male and female sexes are in different flowers of the same plant (the most common sex form).
- *Andromonoecious:* Male and hermaphrodite flowers are in the same plant.
- *Gynomonoecious:* Female and hermaphrodite flowers are in the same plant.
- *Trimonoecious or Androgynomonoecious:* Male, female and hermaphrodite flowers are in the same plant.
- *Dioecious:* Male and female flowers are in different plants. Plants bearing male flowers called androecious and those bearing female flowers called gynoecious plants.
- *Gynodioecious:* In dioecious species some plants of a variety bear only female flowers and others bear hermaphrodite flowers.
- *Sub-gynoecious:* When gynoecious plants of dioecious species produce some male or hermaphrodite flowers.

Monoecious sex form is the most common one in this family. Hermaphroditism is the original sex form from which monoecious and dioecious sex forms have been evolved.

Pollination and Fruit Set

Dioecious and monoecious plants are cross pollinated. Hermaphrodite flowers also fail to set seed by self-pollination some times as the pollen grains are sticky. However, self-pollination between two sexes of the same plant is also common in cucurbits. Poor fruit setting in cucurbits is primarily assumed as the failure of pollination.

However, this situation may be improved by hand pollination. Fruit setting is largely dependent on varietal characteristics, nutrition, disease status and environment. It also depends on the number of fruits already present in a plant. Therefore, harvesting of edible-mature fruits encourages new fruit setting.

Methods of Hybrid Seed Production in Cucurbitaceous Vegetables

Most of the cucurbits are cross-pollinated because of their monoecious flowering behaviour. However, inbreeding may take place without any barrier as both the sexes are present in the monoecious sex form. Experimental results have demonstrated that cucurbits as a whole exhibit no inbreeding depression. It should be pointed out that there are twopopular cucurbitaceous vegetables- Kakrol and pointed gourd (potal), which are of dioecious flowering habit leading to 100 per cent cross-pollination.

These crops may have tremendous inbreeding depression, especially in Kokrol even after one generation of selfing. Inbreeding in dioecious species is not possible in conventional methods as the male and female sexes are in different plants. Moreover, time of anthesis, time and duration of stigma receptivity and pollen availability are different in different cucurbits. Therefore, hybrid seed production technique for each of the cucurbits has been discussed separately.

Different steps of Hybrid Seed Production are:

- Production of inbred lines by inbreeding for 3 to 5 generations.
- Selection of inbred parents through combining ability tests and potential hybrid production ability.
- Production of hybrid seeds (preferably single cross hybrids and pistilate parent preparation is relatively easy, and single fruit produces quite a large number of seeds),
- Maintenance of inbred parents.

WATERMELON

Inbreeding and Selection

Inbreeding may be defined as any system of mating that will lead to an increase in homozygosity. In watermelon, inbreeding is usually practiced by controlled self-pollination. This provides the most effective approach to obtain homozygosity. Inbreeding in watermelon does not cause loss of vigour, self-sterility, decrease in the number of fruits or total yield per plant. Porter stated that inbreeding tends to isolate strains which produce either larger or smaller fruits than the commercial variety. It was also reported by Porter that in

Klondike variety, inbred for four successive generations have been isolated which produce fruits equal in weight to those of the parent variety but excel the latter in fruit uniformity, flesh colour, texture and quality. Single plant selection and compositing the seeds of selected plant after five successive generation will lead the inbred development in watermelon.

Technique of Hybrid Seed Production

Hybrid seed production in watermelon entails little difficulty because of the monoecious nature of the flowers and the large numbers of the seeds contained in a fruit.

Selection of Parental Line

For hybrid seed production, it is very important to choose the parental line for cross combination. The selection of parent should be based on combining ability, cross compatibilitys tudy and hybrid vigour manifested in F1 generation.

At BARI two parental combination *viz.* WM0024 x WM0045 and WM0053 x WM0045 were evaluated based on heterotic performance and cross compatibility and one of the combination was released as F1 PADMA for its light green colour, oblong shape, high sugar percentage and better overall performance.

Maintenance of Parental Line

When a new superior combination of parent lines for an hybrid variety is identified, the first step should be taken on the process of seed production of parent lines by selfing. For the production of seeds of inbreds the following points should be avoided - i) to produce seeds of parent lines every year and ii) to improve the genetical characters of the parent lines by means of producing their seeds early.

Hybrid seeds of watermelon can be produced by two ways.

- *Through artificial pollination.*
- Removal of maleflower and use of insect pollination.

ARTIFICIAL POLLINATION

Field Lay-out:

Seedlings at 3-4 true leaf stage should be transplanted at 1.5 m apart in 2.5 m wide beds. At transplanting time, male parent line should be transplanted in the separate rows, so that they never get mixed with female parent lines.

Selection of Female Flowers

Since watermelon plant bears a number of ineffective female flowers which may fail in fruit set, therefore, effective female flowers have to be selected in order to make successful pollination. Strong and stout female flowers having large ovary with long peduncle are best for effective pollination. The female flower bud which will open next day should be selected for bagging.

Bagging of Female and Male Flowers

The process of bagging are as follows.

- Find out adequate female flowers
- Put the bag covering on the bud
- Insert the bud with vine up to the end of bag and
- Fix the bag on the vine by means of twisting 2-3 times the cut ends of the bag at opposite of the vine

Collection of Male Flowers

There are two methods of collection of male flowers. One is that considering convenience of pollination, male flower buds can be collected in the late afternoon of previous day, keeping their peduncles as long as possible and store them up to the next morning. Another method is to collect the male flowers in early morning before dehiscence of anthers.

Period *of* Pollination

Main pollination period is usually 7 days, 2 days for beginning, 2 days for the peak and another 2-3 days for the last. The peak period of flowering occurs almost in the same dates, therefore, if needed to widen the pollination period, the design of planting date must be adjusted at 15 days interval.

Correlation Between Pollination Hours and Fruit Set

From the mid February to the end of March, flower buds of watermelon starts to open from *6:00* A. M, fully blooms by 700 A. M and completely closes by 11:30 A. M. *On* the other hand during the end of June to the beginning of July, flower bud open from 4:30 A. M, fully blooms by 5:30 A. M and completely closes by 10 A. M.

Pollination

- Collect the male flowers in the box or petri dish.
- Pick the male flower from the box and break the petal, then hold it by the mouth at its peduncle.
- Remove the bag carefully from the opened flower.
- Pick the male flower, hold it at its peduncle and put its anther mass on the pistil tapping all the divided 3-4 stigmas evenly and wholly to pollinate completely.
- Hold the male flower again by the mouth, then put bag on the pollinated flower.
- Put the mark-tag on the vine at the female flowers internode.

In the case of rainy days, the pollinated flowers should have to be protected from rain. Watermelon pollens are very susceptible to water and if touched with water drop the pollens are destroyed immediately.

SEED MANAGEMENT DURING GROWTH PERIOD

When the fruits become the size of baby's head, spread the straws under the fruits in order to avoid phytophthora rots.

Harvesting

Fruits of 30-35 days old from the date of pollination, become ready for harvesting. After harvesting, keep the fruits in well ventilated room for the post-harvest ripening in order to improve the seed colour and for the convenience of seed extraction. When the fruits become over ripe, the seed extraction becomes very easy due to softening of flesh.

Extraction, Washing and Drying of Seeds

It is wise to complete the seed extraction and drying on the same day, therefore, referring the weather forecast and choosing fine day the work is to be started from early morning. The process is as follows. Seeds are mixed with pulp or placenta in cucumber, watermelon etc. Fruits are cut longitudinally into half and scraped out the seed and collected in a barrel. Some placenta remains with the seeds which is to be removed by rubbing with sand or ash followed by washing in water. Otherwise, the pulp surrounding the seeds can be allowed to ferment for 48 hours, when the pulp can be easily separated and then seeds are washed in fresh water.

A rapid method of separating the seed from pulp is by acid treatment. Twenty five to thirty ml of hydrochloric acid or about 8 to 10 ml of commercial sulfuric acid can be used for 5 kg of pulp containing seeds. The seeds can be washed free from the pulp in about 20 to 30 minutes. Then the seeds have to be washed thoroughly to remove excess acid and undeveloped floating seeds are discarded. Seed yield per hectare varies from 150 kg to 300 kg which may vary based on varieties, extent of pollination and field condition.

Removal of male flowers and Use of Insect Pollination

The technique is applicable to monoecious species of the cucurbits, in which male and female parental line should be planted in alternate rows. The male flowers of female plant should be completely removed before their opening. The fruits from the female parent are harvested as crossed fruits and those of the other variety as self fruit of the male variety. This technique has been used in commercial hybrid seed production.In Canada hybrid seeds of muskmelon, cucumber, squash etc. are produced by planting one row of male parent plant and two rows of female parent plants alternately. This methods is referred to as crossing block method. In Japan three workers are required for 0.245 acres to carry out their operation.

PUMPKIN

Production of Inbred Lines

- Seed sowing of diversified genotypes in November to January for winter ecotypes giving 10.6 m distance within and between lines. For summer ecotypes, seed sowing preferably after the 1st monsoon. Allow the plants to climb on narrow bamboo trellis or vertical net-trellis giving 1.0 m interval between plants.
- Bagging of male and female flowers one day before anthesis, and also after pollination of the pistilate flowers for another two days.
- Select the vigorous inbreds after 4-5 generations of inbreeding.

Production of Hybrid Seeds

- Planting of female and male inbred parents in 4: 1 ratio,
- Spray 50 to 100 ml of ethephon per litre of water at 2-3 leaf stage on pistilate parent to increase female flowers at lower nodes.
- Allow the plants to climb on narrow or vertical net-trellis.
- Bagging of male and female flowers every afternoon before anthesis. Rebag the female flowers after pollination for another two days.
- Harvest fruits after about 60 days of pollination.
- Remove seeds from the fruits, wash and dry at low temperature (<30ºC) for 3-4 days and then sun dry for another 3-4 days.
- Preserve the seeds in sealed polyethylene bags at low temperature (4-5°C).

Maintenance of the Inbred Parents

Inbreeding of parents to produce seeds provided the parental stock is depleted. Produce large quantity of seeds to maintain genotypic and phenotypic integrity of the hybrid. Seed Extraction: As in watermelon.

Bottle gourd

In cucurbits, bottle gourd (2n = 22) is second to the pumpkin by area (7300 ha) of cultivation and total production (60,000 t/ year) in Bangladesh. Probable centre of origin is Africa.

Production of Inbred Lines of Diverse Genetic

Bagging of male and female flowers before anthesis. Inbreeding at anthesis through following morning and rebagging of the female flowers for another 2 to 3 days. Inbreeding should be done for 4-5 generations and select the better types by discarding the poor performers.

Production of Hybrid Seeds

Planting of female and pollinator inbreds in 4:1 ratio. Bagging of female and male flowers before anthesis, and only female flowers after pollination for another 2 days. Harvesting of mature fruits after senescence of the plant, remove seeds, wash and dry them. Store in sealed polyethylene bags at low temperature.

Maintenance of The Inbred Parents

Inbreeding of parents to produce seeds should be done when the parental stock is depleted. Produce large quantity of seeds to maintain genotypic and phenotypic integrity of the hybrid.

WHITE GOURD

Area of white gourd ($2n$ = 22) cultivation in Bangladesh is 3,600 ha and total production is 22,000 t/year. Probable centre of origin is Java, Indonesia. Av. yield = 6.11 t/ha.

Production of Hybrid Seeds

Planting of female and pollinator inbreds in 4: 1 ratio. Vertical net-trellis should be allowed for climbing. Bag the female and male flower buds before anthesis and rebag the female flowers after pollination for another 2 days. Harvest mature fruits after 70 to 90 days depending on inbreds. Remove seeds, wash, clean and dry them. Store in sealed polyethylene bag at low temperature.

RIBBED GOURD

Total area under ribbed gourd (2n = 26) cultivation is 4000 ha and total production is 17,000 t/year in Bangladesh. Indian subcontinent is the centre of origin. Av. yield is 4.25 t/ha.

Methods of Hybrid Seed Production

Production of Gynoecious and Monoecious and Moecious/Harmaphrodite Inbred Lines Having Distant Genetic Background. Gynoecious plants should be treated with 300-400 ppm GA3/$AgNO_3$ to induce male sex for selfing to produce gynoecious inbred.

Bagging of male and female flowers in monoecious plants before anthesis. Pollination in the afternoon and rebagging of the female flowers for another 2 days. Inbreeding should be done for 4-5 generations and select the better types by discarding the poor performers. Selection of male and female inbred parents on the basis of their hybrid performances and considering their SCA and GCA value.

Production of Hybrid Seeds

Planting of gynoecious/gynoedioecious/monoecious inbred as female, and monoecious/hermaphrodite/andromo-noecious as male in 4:1 ratio.

Remove hermaphrodite plants from gynodioecious female inbred parent retaining the gynoecious plants only, OR in case of monoecious parents, bag the male and female flowers before pollination and rebag female flowers for another 2 days after pollination. Harvest mature fruits and dry them before seed removal. Seeds should not be stored at low temperature.

Maintenance of The Inbreds

Gynoecious line should be maintained by inducing male sex using 300 - 400 ppm GA3 or $AgNO_3$. Hermaphrodite inbreds can be maintained by simple selfing, and for other sex forms inbred can be maintained in the similar way mentioned for other monoecious species.

SPONGE GOURD

Area of cultivation and total production of sponge gourd (2n=26) in Bangladesh are not known. However, it becomes available in lean period of vegetable supply. Centre of origin is India.

Objectives

High fruit yield having non-bitter taste
Early fruiting habit having late fibre formation in fruit
Resistance against CMV, mites, powdery mildew and downy mildew.

Floral Biology

Anthesis at late night: Another dehiscence in the morning with the rise of temperature. Stigma becomes receptive at anthesis and remains so until noon.

Production of Hybrid Seeds

Planting of female and pollinator inbreds in 4: 1 ratio.
Vertical net-trellis should be allowed for climbing.
Bag the female and male flower buds before anthesis and rebag the female flowers after pollination for another 2 days. Harvest mature fruits after 70 to 90 days depending on inbreds. Remove seeds, wash, clean and dry them. Store in sealed polyethylene bag at low temperature.

SNAKE GOURD

Area of snake gourd *(2n* = 22) cultivation is 2000 ha and total production is 9000 t/year in Bangladesh. Centre of origin is India.

Objectives

High yield with thick mesocarp of the fruits.
Free from bitterness.
Variety with early and late seeding potential.
Floral Biology
Anthesis at or after sunset.
Stigma becomes receptive at anthesis and remains so far at least another 12 hours. Pollination can be done in the following morning of anthesis.

Production of Hybrid Seeds

Planting of female and pollinator inbreds in 4: 1 ratio.

Vertical net-trellis should be allowed for climbing.

Bag the female and male flower buds before anthesis and rebag the female flowers after pollination for another 2 days.

Harvest mature fruits after 70 to 90 days depending on inbreds. Remove seeds, wash, clean and dry them. Store in sealed polyethylene bag at low temperature.

BITTER GOURD

Area of bitter gourd (2n = 22) production is 4000 ha and total yield is 15,000 t/year in Bangladesh. Centre of origin may be Asia, Africa or tropical America. Av. yield = 3.75 t/ha.

Objectives

High yield with at least moderate size of fruit in winter months and long fruits in summer season.

Less bitter taste with thick mesocarp having poor number of small seeds.

Floral Biology

Anthesis at early morning.

Stigma becomes receptive at anthesis and it remains so for another 5-6 hr. Pollen dehiscence starts at anthesis and it increases with the rise of temperature.

Production of Hybrid Seeds

Planting of female and pollinator inbreds in 4: 1 ratio.

Vertical net-trellis should be allowed for climbing.

Bag the female and male flower buds before anthesis and rebag the female flowers after pollination for another 2 days.

Harvest mature fruits after 70 to 90 days depending on inbreds. Remove seeds, wash, clean and dry them. Store in sealed polyethylene bag at low temperature.

KAKROL OR TEASLE GOURD

Exact area of cultivation and total yield is not known. However, large area of cultivation is seen in Brahmanbaria, Akhaura, Rangpur and some other areas. Kakrol (2n = 56) has export potential to Middle East and U. K. Centre of origin probably is Indo-Burma region. It is generally a vegetatively propagated dioecious crop.

Objective

High yield with big size of fruits. Reduced seed number in fruits, and poor root tuberization. Elimination of cumbersome hand pollination practice. Floral Biology Anthesis at early morning. Stigma becomes receptive at anthesis and receptivity remains until mid day. Pollen becomes viable with the rise of temperature, at about 6:00 to 700 A. M.

Method of Hybrid Seed Production

Production of Inbreds of Different Clones on The Basis of Fruit Characteristics. Grow different fruit morphotypes from tubers on vertical trellis of net giving 0.8 m distance between plants in a row. Treat some of the twigs of each clone with 400 ppm AgNo3 before flower bud organogenesis to induce male sex in female plants. Treatment should be given by spraying the solution on top 7-8 leaves of the twigs. Cover the flowers of treated and untreated vines one day before anthesis.

Pollinate the stigma of untreated vines with induced pollen of the same plant or different plants of the same clone. Rebag the pollinated flowers for 3 days to avoid genetic contamination. Collect ripe fruits and dry the seeds for storing at low temperature. Deshelled seeds, without any injury in the embryo, should be sown in moist sterile medium (sand, soil, vermiculite, etc.) maintaining at least 30°C for about 10 days. Seedlings should be established in February every year.

Selection of Inbred Parents

Select two genetically female inbreds, as male and female parents of the hybrid, on the basis of their hybrid performances, SCA and GCA value. Male sex must be induced while producing their hybrids.

Production of Hybrid Seeds

Transplant tuber-originated inbred parents side by side, and allow them to grow on vertical trellis so that pollination can be done conveniently. Induce male sex in at least one of the parents (pollen parent) by spraying $AgNO_3$ (400 ppm) in twigs before flower bud organogenesis.

Bag the flowers of seed and pollen parents 1 day before anthesis. Pollinate the flowers by induced pollen and rebag the pistilate flowers for 3 days. Harvest the hybrid seeds at full maturity and dry them for storing at low temperature. Distribute the hybrid seeds directly or establish the hybrid tubers from F1 seeds and distribute them for commercial production with natural male in 10:1 ratio (female: male) provided there is no parthenocarpy.

Maintenance of The Inbred Parents

Just maintain the seed-originated tubers of the inbred lines and multiply according to necessity. Inbreeding depression may be very high in kakrol. Chance hybrids, though relatively low, from ordinary crosses of male and female genotypes may also be selected and multiplied as variety.

Similar methods of hybrid seed production may be applicable in pointed gourd. However, seed-originated plants take several years for fruit production. Therefore, generation advancement may take time during inbred production.

ONION

Heterosis in Onion

Heterosis in yield of onion hybrids ranged from 14 to 67 per cent when compared with commercially grown onion varieties. Heterosis is manifested in uniform bulb size, bulb weight and an efficient source sink ratio.

Male Sterility in Onion

Male sterility in onion is controlled by the combination of a cytoplasmic factor 'S' together with a recessive nucleus gene in its homozygous form (ms). These male sterile lines are maintained by pollination with a maintainer line of construction 'N ms ms'.

Hybrid Seed Production Techniques

For the production of hybrid seed in onion, the male and cytoplasmic female lines are planted in the ratio of 2 per cent The success of hybrid seed production depends upon the pollen distribution pattern from fertile to sterile plants in the crossing block.

Factors Affecting Hybrid Seeds in Onion

The following factors are responsible for reduced hybrid seed yield in onion.

Weak Inbred lines

Aborteds ovule

Abnormal florets where ovary started to develop but failed to produce seed.

Asynchrony of flowering of parental lines

Excessive heat damaging the flowers.

Seed Yield in Onion can be Improved as Follows:

- Better synchrony of flowering of parental inbreds is achieved by adjusting either storage temperature of the mother bulb (9-14°C) or by planting dates.
- Keeping bee colonies in the hybrid seed production plots @ 3-5 dully developed bee colonies/ha.
- A single application of GA at 50 ppm at the time of first seed stalk emergence reduces the times of 80 per cent of floral stem emergence by half and improves the uniformity of seed stalk height.
- Harvesting seed having 60-70 per cent dry matter content while still in capsule attached to the stalk.
- Avoid shattering of seeds by spraying antishattering materials such as polyvinyl acetate.
- Spray desiccants (diquats) to facilitate uniform drying and mechanical harvesting.

Seed Yields

Hybrid seed yield in onion ranges from 300-350 kg per hectare.

COLE CROPS

Manifestation of Heterosis

Heterosis in cabbage (25-61 per cent), cauliflower (20-60 per cent) and Broccoli (26-58 per cent) is manifested in head/curd size, early maturity, head/curd weight, and plant weight.

Incompatibility and Hybrid Seed Production in Cole Crops

Mild winters (0-5°C) with spring temperature at bloom (15-20°C) are very suitable for cole crop seed production. In cole crops it is the sporophytic incompatibility system which is most prevalent.

The self- and sib-incompatible but cross-compatible lines are set in the field with a planting ratio of one row of pollinator to four rows of seed parent. The lines A (S11) and B (S22) are each propagated by bud pollination or through tissue culture. The hybrid seeds (S1 S2) are harvested from both parents. Such hybrids are the most uniform ones but usually the production cost of the parental lines is prohibitive.

Main problems in hybrid seed production using incompatibility system are:

- Depression by continuous inbreeding of parental lines
- Pseudo compatibility
- Reduction of incompatibility by environmental conditions
- Restriction of pollination within parental lines and
- Matroliny.

Male Sterility and Hybrid Seed Production

Cytoplasmic male sterile lines and their maintainers have been developed in cauliflower, cabbage and broccoli using radish cytoplasm for male sterility. Cytosterile plants of broccoli, cabbage and cauliflower are petaloid with large nectaries responsible for bee attraction and good female fertility.

Seed Yield

It varies from 500 - 800 kg/ha[3]

7

Seed Treatment to Control Diseases

Most seed treatment products are fungicides or insecticides applied to seed before planting. Fungicides are used to control diseases of seeds and seedlings; insecticides are used to control insect pests. Some seed treatment products are sold as combinations of fungicide and insecticide.

Fungicidal seed treatments are used for three reasons:

- To control soil-borne fungal disease organisms that cause seed rots, damping-off, seedling blights and root rot;
- To control fungal pathogens that are surface-borne on the seed, such as those that cause covered smuts of barley and oats, bunt of wheat, black point of cereal grains, and seed-borne safflower rust;
- To control internally seed-borne fungal pathogens such as the loose smut fungi of cereals.

Most fungicidal seed treatments do not control bacterial pathogens and most will not control all types of fungal diseases, so it is important to carefully choose the treatment that provides the best control of the disease organisms present on the seed or potentially present in the soil.

The degree of control will vary with product, rate, environmental conditions and disease organisms present. Some systemic fungicidal seed treatments may also provide protection against early-season infection by leaf diseases. Fungicide-insecticide combination products or an addition of insecticide for wireworm control should be considered if planting newly opened land or land that has had a history of wireworms. Consult current recommendations for insecticides registered for wireworm control. A fungicide-insecticide combination also may be useful for dry beans. The insecticide used on dry beans should be one that provides control of the seed corn maggot.

Seed treatment is a term that describes both products and processes. The usages of specific products and specific techniques can improve the growth environment for the seed, seedlings and young plants. Seed treatment complexity ranges from a basic dressing to coating and pelleting.

In agriculture and horticulture, a seed treatment or seed dressing is a chemical, typically antimicrobial or fungidal, with which seeds are treated (or "dressed") prior to planting. Less frequently insecticides are added. Seed treatments can be an environmentally more friendly way of using pesticides as the amounts used can be very small. It is usual to add colour to make treated seed less attractive to birds if spilt and easier to see and clean up in the case of an accidental spillage.

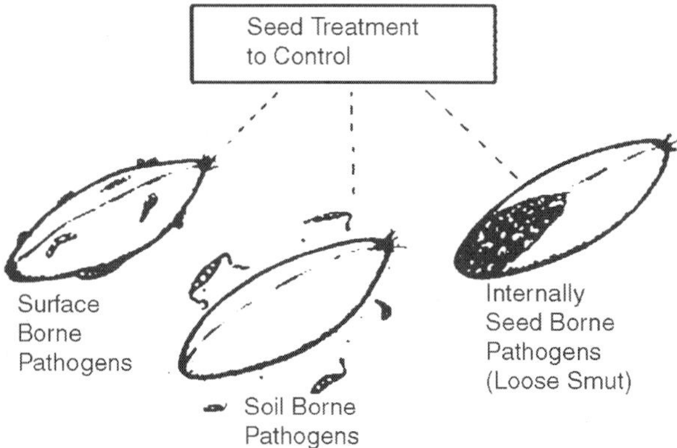

Fig. Reasons for Seed Treatment.

SEED TREATMENT

One seed treatment, imidacloprid, from the neonicotinoid family of insecticides, is controversial and was banned in France for use on maize, due to that government's belief that the chemical was implicated in recent dramatic drops in bee counts, and possibly in the so-called Colony Collapse Disorder. Dust from treated seed is known to have caused at least some problems particularly from crops such as maize drilled during the main honey flows. Improvements to pneumatic drills to reduce dust release, and improvements to seed treatment compounds to prevent the compound breaking up into dust have been introduced in Europe led by Germany and Holland from 2009 to 2012. Information on seed treatments including the information above can be seen on the registration authority databases.

Successful seed treatment is a complex story as it depends on a number of interacting factors. Full compliance with all of them is essential to achieve a high quality seed treatment. It all starts with the seed being treated needing to be clean and of high quality itself. The seed treatment formulation has to be stable and have basic adhesive properties; the film coating has to complete the adhesiveness in order to ensure even distribution of the seed treatment product on the seeds and to provide good seed flowability. The slurry recipe has to be adapted to the seeds, and the volume needs to be adjusted to achieve good coverage and a homogeneous distribution among the seeds. Finally, the equipment must be able to be adjusted precisely and has to work reliably. The treatment will be of high quality only if the operator applies exactly the right dose of both seed treatment product and film coating.

Seed treatment technology has come a long way since the use of salt brine in the mid-1600s. Today, seed treatments deliver clear environmental, economic and social benefits, making the technology a perfect tool for sustainable agriculture.

With a growing world population, now more than ever it's critical that farmers have access to the tools that will help them grow more food while protecting the environment. And seed treatments are one of these tools.

"By utilizing modern agriculture technologies, farmers will be able to boost yields, conserve water usage and protect biodiversity," says Keith Jones, director of stewardship and sustainable agriculture for CropLife International based in Brussels, Belgium. "Seed treatments represent one tool on which many sustainable agriculture technologies rely upon. By protecting seeds from planting to emergence, seed treatments can improve stand establishment and increase potential yield."

Helmut Schramm, head of the seed treatment business at Bayer CropScience in Monheim, Germany, agrees, saying seed treatments deliver clear environmental, economic and social benefits, making this technology a perfect tool for sustainable agriculture.

"Innovative seed treatment technology represents an environmentally-sound approach to crop protection," he says. "Treating the seed provides a targeted and effective means of application that helps increase yields, safeguard our environment and ensure a sustainable means of crop production."

Over the years, seed treatments have evolved from simply protecting the seed to helping improve plant stand and early plant health.

"Seed treatments are increasingly designed to also enhance plant emergence, growth or nutrition efficiency, which, along with crop protection, lead to a more vigorous and uniform crop," says Schramm. "This forms the optimal base for a high-quality crop which can fully exploit its yield potential."

Greg Lamka, chair of the International Seed Federation Seed Treatment and Environment Committee, says that in order to maximize yield, it's important to have a full and uniform stand. The STEC committee was established in the 1990s to raise the seed industry's awareness about the use of different seed treatments and to promote a better understanding of how production could be improved and made more efficient. The committee consists of seed companies and crops protection companies who wish to promote the safe and effective use of seed treatment products. "Growers are paying more for the seed, so expectations are rising about the performance of our products. Seed treatments are a way to ensure that the products will perform to their maximum, based on the environment that they're put into," says Lamka, adding the vast majority of the seed used in developed countries is treated.

ACID TREATMENT

The Chemical most commonly used to break seedcoat dormancy is concentrated sulfuric acid. For some species it is more effective than hot water treatment. Seed which has been kept for a long period in store may require a longer period in the acid than fresh seed, which could be severely damaged by the same length of treatment (Kemp 1975c). Great care is needed in the handling of sulfuric acid and this method is not suitable for use by unskilled workers. Detailed instructions for use of sulfuric acid are given by Bonner *et al.* 1974 and are reproduced below:

Materials and equipment required are as follows:

- Commercial grade (specific gravity 1.84, 95 per cent pure) sulfuric acid; acid-resistant containers (thick plastic preferred); wire containers and screens for handling, draining, and washing the seeds; an abundant supply of running water; a safe place to drain the dilute acid resulting from rinsing the seeds; and facilities for drying the seeds after rinsing.
- Safety Precautions are a must! All workmen must understand and obey safety precautions in the use of acid.
- Seeds, containers, implements, and the acid itself must be handled with great care to avoid injury.
- Water must not be splashed into the acid, as a violent reaction will occur.
- All workmen should wear suitable safety clothing, gloves, and goggles or other eye protection.

Toughness of the seedcoat varies between lots and even between individual trees in most species. The optimum period of immersion in acid for each lot may be determined by treating a small sample for different periods and then soaking the lots in water at room temperature for 1 to 5 days (depending on species). The treatment period that yields a high percentage of swollen seeds (by water uptake) without visible injury is the right one. Oversoaking may pit the seed and even expose the endosperm. Insufficient soaking leaves the seedcoats of most species glossy; coats of correctly treated seeds are dull, but not deeply pitted. If Tests reveal only small differences between lots, then all may be lumped together for treatment, unless there are other reasons for keeping them separate (such as seed source distinctions). Large differences between lots should occasion separate treatment.

The steps in acid treatment are as follows:

- Allow seeds to come to air temperature. If they have been removed from cold storage, do not open the container until temperature equilibrium is reached. Moisture will form on cold seeds exposed to warm moist air, and this moisture can react with acid to raise temperature to the danger point.
- Thoroughly mix seeds to be treated as one lot.
- Immerse seeds in the acid for the required period, making sure that all are covered. Treatment should be carried out at 65° to 80 °F (18°–27 °C), preferably on the upper end of the range (Heit 1967a). Lower temperatures require longer soaking times than do higher temperatures. Careful stirring will reduce the length of treatment necessary.
- Remove seeds from the acid and wash them promptly and thoroughly in cool, running water for 5 to 10 minutes to remove all traces of acid. Water should be applied copiously at the start, and the seeds stirred carefully during rinsing.

BIOLOGICAL METHODS

In Nature animals and micro-organisms are an important factor in the breakdown of seedcoat impermeability. It is difficult to make use of these organisms as a controlled pretreatment of seed, but in a few cases successful results have been obtained. Seeds of *Acacia senegal* and *Ceratonia siliqua* that have passed through the digestive tracts of goats germinate readily when placed in favourable conditions, because of the action of the strong digestive chemicals. Feeding the pods to penned goats and collecting the seeds from the droppings is a convenient pretreatment for these species (Goor and Barney 1976).

Seeds of some species are said to be regurgitated after partial rumination *e.g. Gmelina arborea* (Greaves 1981). Troup (1921) states that seeds of *Acacia nilotica* are ejected after rumination by sheep and goats but pass right through the digestive tract in cattle. In either case germination is improved by the digestive action.

Termites are an important agent for breaking down seedcoat dormancy in many parts of the tropics. In Thailand teak fruits were spread on the ground in a 5 cm thick layer immediately after collection and covered with cardboard. After about 5 weeks the termites had removed the exocarp and subsequent germination, after alternate wetting and drying, was significantly improved in comparison with fruits sown with intact exocarps (Bryndum 1966, Sompherm 1975). Termites have been used in a similar way to break down the tough winged and bristly fruit of *Pterocarpus angolensis* (Groome *et al.* 1957). Periodic inspection is essential to ensure that the process is not carried too far.

Partial Fermentation, which is damaging to many seeds, can be beneficial in overcoming seedcoat dormancy in some. In the southern Sudan fruits of Tectona grandis are allowed to lie on the ground through the rainy season for partial fermentation. They are then collected, stratified in a pit with layers of,

- Seed
- Organic matter
- Soil and watered daily for 10 days.

Satisfactory germination results (Wunder 1966).

DRY HEAT AND FIRE

Solar Radiation is not used alone to promote germination but is an important component of the alternate soaking and drying treatment described above under "Soaking in water". In the Seasonal wet and dry tropics, fire is a powerful natural factor in the removal of seedcoat dormancy. A fierce fire will kill the seeds but a light to moderate fire, such as those associated with controlled early burning, will reduce seedcoat impermeability and stimulate germination. Fire has been used in a number of countries to stimulate germination of *Tectona* (Laurie 1974). The fruits may be spread thickly on the ground and covered with grass which is burnt off, or they may be lightly scorched by a flame gun. Adjusting the heat of the fire to achieve the maximum effect on the pericarp without damaging the seed embryo requires experience.

Similar treatment is used for *Aleurites moluccana* in the Philippines. The nuts are spread evenly on the ground and covered with a 3 cm thick layer of dry *Imperata* grass which is set on fire. As soon as the grass is burned, the seeds are placed in cold water. The quick change of temperature causes the nuts to crack and they are ready for sowing (Seeber and Agpaoa 1976). An alternative is to sow the nuts at correct spacing with only half their diameter in the soil. A layer of *Imperata* grass is spread over the seed bed and set on fire. After burning, the seedbed is sprinkled with water and the nuts are pushed 2 cm deep into the soil and watered thoroughly.

In Sabah Bowen and Eusebio (1981b) found that ten minutes' exposure of *Acacia mangium* seeds to dry heat at 100° C was nearly as effective in overcoming dormancy as the immersion in water at 100 °C. Germination was 83 per cent compared with 92 per cent in the hot water treatment.

CHEMICAL TREATMENT OF PHYSIOLOGICAL DORMANCY

A Wide range of chemicals have been tested experimentally in an attempt to overcome internal dormancy. They include gibberellic acid, citric acid, hydrogen peroxide and a number of other compounds. Some have given a degree of improvement, *e.g.* Bachelard (1967) found that the germination of dormant seeds of *Eucalyptus delegatensis, E. fastigata* and *E. regnans* could be improved by treatment with gibberellic acid (GA). 24 hours' immersion in either GA 3 or GA 4/7 of *Nothofagus obliqua* has given rapid and complete germination in 14 days, although this normally dormant species otherwise requires 28–42 days' stratification (Gordon 1979). Shafiq (1980) found that the strength of the gibberellic acid had only a small effect, 200 ppm giving 100 per cent germination in 8 days, 50 ppm the same germination in 12 days.

The best stratification treatment (42 days at 3–5 °C) yielded 70 per cent in 14 days and 88 per cent in 28 days and the control (24 hours' soaking in distilled water and no prechilling) only 20 per cent in 28 days. The saving in time effected by the GA treatment (1 + 12 days compared with 42 + 28 days) is considerable. Results reported later by Rowe and Gordon (1981) showed that GA 4/7 was more reliable than GA 3 as it was less sensitive to temperature during the germination period. Excellent germination was obtained throughout the temperature range 15–30 °C, whereas GA 3 required temperatures over 21 °C for comparable results.

These successes, however, are the exception. In general, for cheapness and reliability chemical treatments cannot compete with stratification or moist prechilling and they are unlikely to play a major role in normal nursery practice in the foreseeable future.

TREATMENTS FOR ENDOGENOUS DORMANCY

X-rays, Gamma rays, light rays in the red region of the spectrum and high frequency sound waves have all been used experimentally to try to overcome dormancy and stimulate germination. Improvement has been reported in some species including *Tectona* (Bhumibhamon 1973), but it has proved difficult to achieve consistent results and the treatments may induce chromosome damage and other abnormalities (Kemp 1975c). Lynn (1967) concluded that there was far more evidence of unfavourable effects from irradiation of seed than beneficial effects. None of these methods are suitable for practical application at the present time.

MOIST PRECHILLING METHODS

Similar Results to stratification in layers may be obtained with many species by storing the seeds moist in polythene bags. As with indoor stratification, seeds should be soaked in several times their volume of water before prechilling, a 48 hour soak at 3–5 °C is suitable for many temperate broadleaved species (Gordon and Rowe 1982). After soaking, the water is drained off and the moist seeds are then prechilled at 3–5 °C for the period appropriate to each species.

Prechilling may be "naked" *i.e.* without any medium, or the seed may be mixed with 2–4 times its volume of a medium such as moist sand, moist peat or a mixture of the two Polythene bags of about 100 micron thickness make suitable containers since they are moisture proof but somewhat permeable to oxygen. They should be lightly tied and opened weekly when the seed should be mixed and, if necessary, remoistened. A smell of alcohol on opening a bag indicates that anaerobic respiration is taking place because of inadequate oxygen; in this case frequency of opening and mixing should be increased (Bonner *et al.* 1974).

Naked prechilling has the advantage that it is easier to check the condition of the seeds throughout the prechilling operation and there is no need to separate seeds from medium at thve end of treatment. On the other hand, there is evidence that germination in some species benefits from use of a medium. Gordon and Rowe (1982) reported less than 30 per cent germination in 50 days from seeds of *Sambucus racemosa* treated "naked", compared with 60 per cent in 20 days from seeds in peat/soil mixture, other elements in the treatments being identical. These authors give detailed prescriptions for pretreatment of a large number of temperature broadleaved species; as a general rule "naked" prechilling is satisfactory for species which need only a few weeks' prechilling, while use of a medium is advisable for those which need a longer prechilling period and for all species which need warm moist pretreatment.

Periods of prechilling vary considerably from species to species and, to some extent, from seed lot to seed lot within a species. For *Abies* a period of 3 weeks at 3–5 °C has proved satisfactory (Aldhous 1972). The same temperature and period is effective for most of the cold-temperate eucalypts, but some provenances of *E. delegatensis* need 4–8 weeks for fast, uniform germination. For *Nothofagus obliqua* and *N. procera* naked prechilling at 3°–5 °C for six weeks, and surface drying before sowing, gave high germination (usually over 80 per cent in 28 days) under nursery conditions (Rowe and Gordon 1981). But, as described later in this chapter, treatment with gibberellic acid was a reliable and simpler alternative. In species having deep physiological dormancy the prechilling period may be as long as 20 weeks *e.g.* Liriodendron tulipifera(Bonner *et al.* 1974, Gordon and Rowe 1982).

For Fagus sylvatica in poland, after storage for several years at –5 °C and 10 per cent MC, the following pretreatments are recommended (Suszka 1979, Suszka and Kluczynska 1980):

1. Allow to Defrost.
2. Moisten by sprinkling of water and thorough mixing of wetted nuts twice a day for 6 days, at 3 °C, until MC rises to 31 per cent.
3. Leave wet nuts in unsealed containers without any storage medium at 3 °C for a period equal to two weeks longer than the minimum required to produce 10 per cent germination in a sample within two weeks of its transfer to a moist germinating medium. Recurrent sampling and germination testing is needed to estimate this period, which may vary considerably from seed lot to seed lot. MC of 31 per cent to be maintained meanwhile by periodic weighing of containers and remoistening of nuts to restore loss of weight.

4. Sow in moist germinating medium at 3 °C and leave for two weeks. This should initiate germination of radicles.
5. Transfer to temperature of 20 °C to promote hypocotyl and epicotyl elongation and seedling emergence, which are inhibited in this species at 3 °C.

For Large-scale operational sowings, the ideal experimental conditions described above can be simulated to some extent in the nursery by timing the spring sowing so that the nuts experience an adequate period of cold temperature first, followed by the higher temperatures of late spring and early summer.There is Some evidence that Steps (2) And (3) Of the pretreatment described above can be carried out before, as well as after, storage. The advantage is that the nuts are ready for sowing as soon as they come out of store, without the need for a subsequent cold, moist treatment covering several weeks. Nuts pretreated before storage have been successfully stored for 15 months in France (Muller 1982).

TREATMENTS DESIGNED TO OVERCOME DOUBLE DORMANCY

Some species combine more than one form of dormancy at the same time. Pretreatment to break one type of dormancy alone will be largely ineffective, unless it is followed by a second pre-treatment to overcome the other type.

Physical dormancy of the seedcoat may be combined with physiological dormancy of the embryo. In this case seedcoats should be treated first *e.g.* by scarification and cold moist prechilling applied afterwards to overcome embryo dormancy. *Cercis canadensis* provides a good example; single treatments gave less than 10 per cent germination in all cases, while successive applications of seedcoat and internal dormancy treatments gave rapid germination of 45 per cent (mechanical scarification) or 65 per cent (acid scarification) (Bonner *et al.* 1974).

In a Few cases treatment of the seed covering after stratification has proved effective. Recent trials with *Fraxinus pennsylvanica* showed that cold stratification at 4 °C for 88 days produced 35 per cent germination in 3 weeks, compared with 2 per cent in unstratified seed. But removal of the pericarp after stratification further increased germination to 56 per cent, while the best, but laborious, treatment of stratification followed by removal of embryos from pericarp and endosperm gave 88 per cent (Marshall 1981). Treatment to increase permeability of the pericarp in acorns after stratification benefits germination.

Tuskan and Blanche (1980) found that 6 hours of mechanical shaking in distilled water, after 3 months' storage at 0 °C and 30–40 per cent MC gave 90 per cent germination in *Quercus shumardii,* compared with 70 per cent from 6 hours' simple soaking after storage. The shaking treatment also increased MC over that from the soaking treatment, indicating increased permeability.

Fraxinus excelsior combines morphological dormancy (underdeveloped embryo) with physiological dormancy. Warm moist treatment to remove the morphological dormancy should be followed by cold moist treatment to remove the physiological dormancy. For UK conditions Gordon and Rowe (1982) recommend 8–12 weeks warm, followed by 8–12 weeks cold, while in Poland, Suszka (1978 a) recommends 16 weeks of each treatment.

A Number of species, especially in the family Rosaceae, combine mechanical dormancy, caused by a tough, thick pericarp, with physiological dormancy. The same combination of warm moist followed by cold moist pretreatment is effective. In the case of a resistant species, *Crataegus monogyna*, the recommended periods are 4–8 weeks warm, followed by 12–16 weeks cold (Gordon and Rowe 1982).

In some species, germination of the radicle takes place readily at warm temperature, but the epicotyl will not start to grow until,

- The radicle has already started to germinate and
- The seed has been exposed to a period of low temperature.

Examples are *Viburnum opulus* and *Carpinus caroliniana* (Bonner *et al.* 1974). Here again the same sequence of warm moist treatment at 20–25 °C followed by cold moist treatment at 3–5 °C should induce satisfactory germination.

Thus the same combination of warm moist plus cold moist treatments applied in that order is capable of removing several different combinations of double dormancy.

SPECIAL TREATMENTS FOR MECHANICAL DORMANCY

The Thick, tough but water-permeable coverings of seeds exhibiting mechanical dormancy prevent embryo growth even when water can be freely imbibed. This mechanical obstruction to germination can be overcome by a period of "warm moist" treatment the length of which varies according to species.

The treatment recommended by Gordon and Rowe (1982) for temperate species is:

1. Soak Seeds in several times their volume of cold water at approximately 3–5 °C for 48 hours.
2. Drain off the water and mix the seeds with two to four times their volume of a moistened, water-retaining medium such as sand, sand/peat mixture, vermiculite.
3. Store at a warm temperature. A constant 20–25 °C or alternating 20° and 30 °C is suitable for many species.
4. Open containers weekly, mix seeds and, if surfaces show signs of drying out, remoisten with water spray.

The period of "warm moist" treatment can be shortened in some species by a preliminary treatment in sulfuric acid. This requires more care and expertise than the use of acid for physical dormancy. Seeds or fruits must be thoroughly dried before treatment and the process should be limited to the partial digestion of the outer layers only, leaving the final weakening of the inner layers to be done by the subsequent warm moist treatment (Gordon and Rowe 1982). In most cases it is preferable to accept the safer but slower method of warm moist treatment alone. Periods of treatment vary from 2 weeks for some species of *Prunus* to 16 weeks for some species of *Crataegus*.

At the end of the appropriate period seeds which possess only mechanical dormancy are ready for sowing. Many species in this class also have physiological dormancy of the embryo.

It May be noted that the warm moist treatment which removes mechanical dormancy is identical with the warm moist treatment which removes morphological dormancy (underdeveloped embryos).

SEED TREATMENT APPLICATION

Fungicide seed treatment products come in a variety of formulations and in a variety of packaging sizes and types. Some are registered for use only by commercial applicators using closed application systems, others are readily available for on-farm use as dusts, slurries, water soluble bags, or liquid ready-to-use-formulations. Whatever the formulation used or application method chosen, some precautions should be taken to assure applicator safety and appropriate seed coverage.

CAUTIONS

Follow label directions when handling seed treatment chemicals. These products are potentially poisonous if mishandled or misused. Extreme caution must be used when handling seed treatment chemicals: some are toxic, others may be irritating. An approved chemical respirator and goggles are recommended even if not specifically required by the fungicide label. The rate of application prescribed by the label must be used: over treatment may injure the seed and under treatment may not provide good disease control. To apply the correct rate, it is essential to calibrate application equipment carefully and to check calibration frequently.

Metering cups of commercial applicators should be cleaned daily to prevent a buildup of chemical that might result in reduced application rates. An auger which has been used to treat seed cannot be cleaned up sufficiently for use in augering grain for food or feed. Once an auger is used for seed treatment, it should be used only for treatment or augering seed for planting. It should not be used to auger grain used for food or feed. Treated seed should not be used for food or feed, and treated grain should not contaminate grain delivered to elevators or be placed in bins or in trucks delivering to elevators.

Containers should be triple rinsed with the rinse water added to the treatment mixture. The rinsed containers should be punctured and crushed for disposal in an approved landfill.

CEREAL SEED TREATMENTS

Fungicidal seed treatments help control soil-borne pathogens that cause seed decay, seedling blight and root rot. Control of these diseases may result in better stands, more vigorous seedlings, and increased yields. Protectant fungicides such as captan, maneb, PCNB, thiram, or fludioxonil (Maxim) help control most types of soil-borne pathogens, except for common root rot and take-all.

Protectant fungicides containing captan, maneb, or thiram are sold under various trade names. Fungicidal seed treatments to protect against the soil-borne fungi that cause common root rot and take all will be discussed under barley seed treatment.

Fungicidal seed treatment controls most, but not all, of the seed-borne diseases of small grains. Seed treatment should be used when the seed is contaminated with smut, scab, or black point fungi. Specific recommendations for these diseases are discussed under barley and wheat. No seed treatment is a substitute for good seed.

Barley

Barley has three smuts: covered smut, black semi-loose smut (nigra smut), and loose smut. Covered smut and black semi-loose smut are surface-borne fungal pathogens that infect the emerging seedling. These two smuts can be controlled by various protectant fungicides. Loose smut infects the embryo of the seed before harvest.

Protectant fungicides do not control loose smut—only the systemic fungicides carboxin (Vitavax or Enhance), triadim- enol (Baytan) or tebuconazole (Raxil) will control loose smut in barley. These same products also will control the covered and semi-loose smuts of barley.

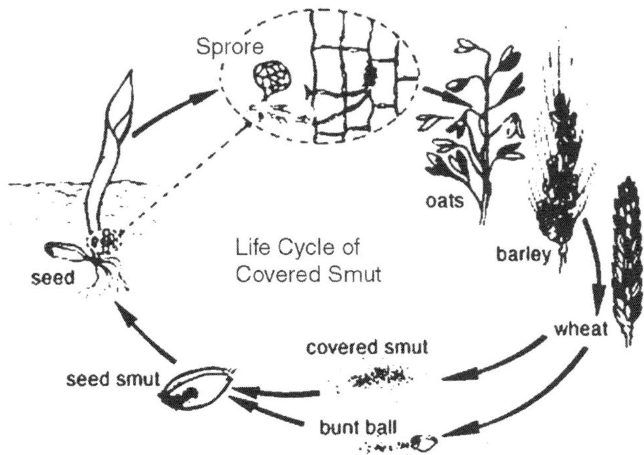

Fig. Life cycle of covered smut and bunt.

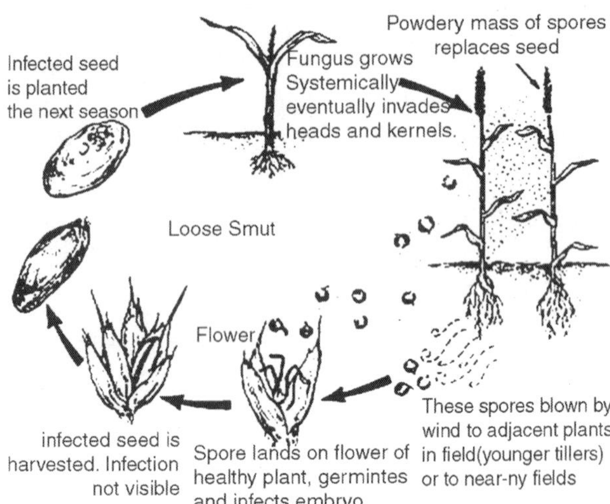

Fig. Life cycle of loose smut.

Covered Smut

Seed of barley varieties susceptible to covered smut should be treated with a protectant fungicide or with carboxin, triadimenol or tebuconazole. At present, all barley varieties recommended for production in North Dakota must be considered susceptible.

Black Semi-loose Smut

Many varieties are susceptible to this smut. It is indistinguishable from loose smut in the field, but it is borne on the seed surface and infects seedlings just as covered smut does. It is not detected by the embryo test, because it does not infect the embryo. It is controlled by seed treatment with a protectant fungicide or with carboxin, triadimenol, or tebuconazole.

Loose Smut

Many of the barley varieties released in the late 1970s and early 1980s were resistant to the races of the loose smut fungus present at that time. In the mid-1980s, a new race of the loose smut fungus was detected in North Dakota. Seed treatment for loose smut in barley planted for seed production is especially important. Seed samples of susceptible barley varieties should be sent to the North Dakota State Seed Department for an embryo test to determine the percentage of loose smut infection.

Losses from loose smut are about equal to the per cent infection; 5 per cent loose smut represents a yield loss of about 4.3 per cent. If the embryo test shows 2 per cent or greater infection, the seed should be treated with carboxin, triadimenol or tebuconazole, or loose smut-free seed should be used for planting.

Common Root Rot

The fungus that causes common root rot is soil-borne and is widespread in North Dakota soils. The common root rot fungus increases under barley or wheat culture, and it often causes mature plant root rot. The fungus also may cause severe root infection in the seedling stage and if splashed to the leaves and head causes spot blotch and black-point infections, respectively.

A number of seed treatment fungicides are now labeled for wheat and barley for suppression of mature plant root rot and also for suppression of seedling blight due to the common root rot fungus. Imazalil is available under several trade names. Triadimenol (Baytan) also is registered for suppression of seedling blight due to common root rot. Tebuconazole (Raxil) is registered for wheat and barley, and difenoconazole (Dividend) is registered only for wheat at the time of publication of this circular. The conditions in which seed treatment for common root rot would be most beneficial are: where continuous wheat or barley is grown, or if short rotations between these susceptible crops are practiced, and in soils where moisture stress is very likely.

Barley Stripe

This fungus disease is rarely seen in North Dakota but occasionally appears in two-row barley if contaminated seed is planted. The disease causes distinct yellow, then brown stripes that run the entire length of the leaves. The stripes join together and the leaves become split and frayed or shredded in appearance.

The fungus is seed-borne in the hull and seed coat. If seed from a suspected disease source is used, seed treatment with a carboxin + thiram or tebuconazole product provides some control of the barley stripe fungus. The systemic fungicide imazalil provides good control of barley stripe.

CORN

Corn seed is especially susceptible to attack by soil-borne pathogens when sown in cold wet soil, when the seed is in poor condition, when it is mechanically injured, or if it has been stored for two years or more. Seed treatment will protect against seed rot and reduce the danger of seedling blight. Sweet corn is more susceptible to attack than field corn, but both should be treated. Most field corn is already treated when purchased. A number of protectant and systemic fungicides are registered for control of seedling blights and seed rot in corn.

WHEAT

Bunt or Stinking Smut

The fungus causing this disease adheres to the seed surface and then infects the emerging wheat seedling. The bunt fungus has a fishy odor and imparts the same odor to flour made from bunted kernels. The price of bunted wheat is discounted for this reason. Bunt is not known to occur in North Dakota at the present time, nor has it occurred in recent years. However, it occurs in other wheat-producing states in the Great Plains.

Growers who purchase seed from out of state should use a seed treatment fungicide, since out-of-state seed could be infested with (carrying) bunt. Growers who have their crop custom combined and then save their own seed should use a fungicide, since the combine could be contaminated with bunt spores or bunt balls from states to the south. Many protectant, systemic, or combination products are available to control bunt.

Loose Smut

Loose smut of wheat infects the embryo, as with loose smut of barley. However, the wheat loose smut pathogen does not infect barley and the barley loose smut pathogen does not infect wheat. No embryo test is available for detection of loose smut in wheat.

Counts of loose smut-infected heads can be made in wheat and durum fields to estimate the per cent infection in the crop. Counts should be made at flowering time, as loose smut heads are hard to detect later. The counts provide information on the per cent infection in the seed that was planted but are not a reliable estimate of the amount of infection in the seed of the harvested crop.

If weather conditions are favorable, *i.e.* cool and wet, infection in the harvested seed could be considerably higher than in the seed that was planted. Carboxin, difenoconazole, tebuconazole, and triadimenol seed treatments control wheat loose smut.

Scab and Black-point

Scab and black-point are fungal diseases that attack both wheat and barley seed. The scab fungus also may infect oats. The scab fungus infects the kernels during flowering if warm, wet conditions prevail. Black-point infection occurs from heading to maturity and also is favored by warm, wet weather.

Scabby seed is shriveled, light in test weight, and often has a chalky white or pink discolouration. When planted, scabby seed has poor germination and poor vigour. The scab fungus does not grow systemically from the seed through the plant to cause subsequent head scab infection. The source of head infection is spores from infected small grain or corn residue. Black-pointed seed has a black to brown discolouration of the embryo or germ end, a discolouration which may extend around the kernel into the crease. When planted, black-pointed seed also may have low germination; if the seed germinates, the seedling roots are often infected by the fungus.

Protectant seed treatments or protectants in combination with systemics, or systemic products alone have provided significant improvement in stand and vigour. Yield increases also have occurred with treatment of scabby and black-pointed grain. Seed treatment helps assure good stands and seedling vigour; it will not prevent possible scab or black-point infections of the head during the growing season.

Take-all

Take-all is a root disease of barley and wheat caused by a fungus that thrives in very wet soils. Take-all generally is not a serious or common disease in barley or wheat in North Dakota, unless the crop is grown under irrigation. Triadimenol (Baytan), difenoconazole (Dividend) and tebuconazole (Raxil) are registered for suppression of this disease.

PYTHIUM

Damage to wheat from the *Pythium* fungus has not been well documented in North Dakota, but the fungus is common in agricultural soils and generally does more damage where wheat is planted into crop residues and soil temperatures are cool and soil moisture is high. Several seed treatment fungicides are registered for control of *Pythium* spp. in wheat, including difenoconazole + mefenoxam (Dividend XL), mefenoxam (Apron XL), and metalaxyl (Allegiance).

OATS

Smuts on oats have not been reported as a recent problem in North Dakota, although they have been serious in other states in the region. Oat smut can cause severe losses. Distinguishing between covered and loose smuts of oats in the field is not easy.

Oat loose smut is more difficult to control with protectant fungicides than oat covered smut, because the loose smut spores are lodged under the hulls where they are difficult to reach. Carboxin and tebuconazole are registered for oat loose smut control.

LEGUMES

Alfalfa and Small Seeded Legumes

Alfalfa, clover and other small seeded legumes may be treated for seedling blight diseases and damping off. The extent of occurrence of these disease problems in North Dakota has not been documented.

Captan or thiram products are registered for seedling blight control, while mefonaxam, metalaxyl, and oxadixyl products are registered for control of *Pythium* damping off and early season infections by *Phythophthora.*

Soybeans

Generally, seed treatment is not required for soybeans. However, it may pay to treat seed if: the germination is below 85 per cent; the seeds are contaminated with fungi; the seeds are badly weathered or injured; or the seed coats are broken. If captan or PCNB treated seeds are to be inoculated with *Rhizobium* bacteria, the inoculant should be applied in-furrow.

Thiram, fludioxonil, metalaxyl and mefenoxam have little or no effect on *Rhizobium* bacteria. Generally, inoculant is not used on soybeans planted on land that was previously cropped to soybeans.

In fields with *Rhizoctonia* problems, a seed treatment containing carboxin, chloroneb, or PCNB should be used. Seed infested with Sclerotinia should be treated with fludioxonil, which will control infection of seedlings by the Sclerotinia fungus.

Dry Edible Beans

Most dry bean seed is treated prior to sale, often with a combination of fungicide, insecticide and bactericide. A fungicide protects against seed- and soil-borne fungi. The bactericide streptomycin controls surface-borne blight bacteria but will not control internally-borne blight bacteria.

Chickpeas

Several seed treatment fungicides are registered for use on chickpeas. Fludioxonil (Maxim) gives broad spectrum protection against soil-borne fungi; mefenoxam and metalaxyl (Allegiance) provide very specific and highly effective protection against *Pythium* seed rot and seedling blight and early season *Phytophthora* root rot.

Chickpea growers in some parts of western North Dakota have experienced severe difficulty with seed rot of untreated seed and should treat seed with mefanoxam or metalaxyl.

The most serious foliar disease of garbanzo beans is *Ascochyta* blight. *Ascochyta* can be seed-borne, usually at low levels. Wet weather may cause rapid spread of *Ascochyta* from a few infection centers. Thiabendazole (TBZ) is registered for use on garbanzo beans for eliminating seed-borne *Ascochyta*. Purchase of western grown seed that has been laboratory checked for *Ascochyta* also is desirable to minimize the danger of losses from this serious disease, but it should also be treated with thiabendazole.

Lentils

Lentils should be treated with captan, fludioxonil (Maxim), mefenoxam (Apron XL) or metalaxyl (Allegiance). Captan and fludioxonil provide broad spectrum protection against *Rhizoctonia* and *Fusarium* seedling blights. Mefenoxam and metalaxyl (Allegiance) provide excellent control against *Pythium* seedling blight, but do not provide protection against *Rhizoctonia* or *Fusarium*.

The two most serious diseases of lentils are anthrac-nose and *Ascochyta* blight. Both can be seed-borne at low levels but seed transmission to developing seedlings has not been deomonstrated for anthracnose. Ascochyta can spread rapidly in wet weather from a few infection centers. The *Ascochyta* that infects lentils is a different species than the one that infects chickpeas. No fungicide currently registered is effective for elimination of either of these disease fungi on the seed. Purchase of western grown seed that has been laboratory tested for *Ascochyta* is desirable to help minimize the danger of losses.

Flax

Flax seed can be attacked by seed- and soil-borne pathogens, especially when the weather is unfavorable for germination and growth or when the seed coats are cracked and split. Seed treatment with protectant fungicides captan, mancozeb, maneb or thiram will reduce the amount of seed rot and seedling blight.

Yellow-seeded varieties are more susceptible to seed coat damage than are brown-seeded varieties. All varieties currently being grown are brown-seeded varieties except Omega, which is yellow-seeded.

Peas

Common seedling blights of peas can be controlled with captan, fludioxonil, PCNB or thiram. The water mold fungi *Pythium* and *Phytopthora* are common problems on peas and can be controlled with mefenoxam, metalaxyl or oxydixyl treatments. No seed treatment is registered for Aschochyta blight on peas, but data from Manitoba indicates that thiram provides good suppression of seed-borne Ascochyta.

Safflower

Safflower rust is borne on the surface of the seed and produces infections on the hypocotyl of the emerging seedlings. Seed-borne rust can result in poor stands and reduced vigour of the seedlings. Fungicidal seed treatment of safflower with one of several available fungicides is recommended to control seed-borne safflower rust. Carboxin, mancozeb, and thiram are labeled for seed treatment use on safflower.

The winter spores of safflower rust can survive from one season to the next in the soil but will not survive to the second season. Control of safflower rust requires seed treatment with a fungicide and crop rotation. Never plant safflower on land that had safflower the year before.

Sugarbeets

Pythium, Aphanomyces, and *Rhizoctonia* are fungi that may cause stand establishment and seedling disease problems in sugarbeets. The *Pythium* fungus occurs in most sugarbeet soils. It causes seed rot, pre-emergence damping off, and post-emergence damping off.

Post-emergence damping off caused by *Pythium* may occur when the seedlings are so tiny that they dry up and blow away within a day or two.

Consequently, Pythium-induced seedling death is seldom noticed by the grower. The only thing noticeable may be an unusually poor emergence and stand. *Aphanomyces* and *Rhizoctonia* cause death of seedlings at a later growth stage, with plants progressively dying from the two to the eight leaf stage. Later in the season, both fungi may cause a root rot which weakens the plant and also reduces the weight and quality of beet roots. Both diseases are favored by warm soil conditions. *Aphanomyces* is favored by heavy soils with poor drainage, resulting in saturated or puddled soils. *Rhizoctonia* is favored by moist soils.

Most sugarbeet seed is sold treated, but different treatments vary in their effectiveness against these three fungi. Certain fungicides have specific activity: mefenoxam, metalaxyl (Allegiance) and oxadixyl (Anchor) are highly effective against *Pythium;* thiram is moderately effective against *Pythium;* PCNB, chloroneb and fludioxanil (Maxim) are effective against *Rhizoctonia.*

Growers planting in fields with known *Rhizoctonia* problems may wish to request a special or supplemental seed treatment from their seed supplier or else use a planter box overtreatment. Hymexazol (Tachigaren) is effective against *Aphanomyces* and *Pythium.*

Canola

Blackleg is a fungus disease of canola that can cause severe losses. It is seed-borne and also spread by wind-borne as well as rain-splashed spores. The wind-borne spores come from canola crop refuse. Most long distance spread of the blackleg fungus is on seed. Most spread within a field or between fields is from wind-borne spores. The severe (highly virulent) strain of blackleg is common.

Benomyl seed treatment provides excellent control of the seed-borne phase of blackleg. Registration of other seed treatments effective against blackleg may occur in 2000 or 2001. Seed treatment is highly recommended for all canola seed planted in areas that do not yet have the severe strain of blackleg. In areas where the severe strain is prevalent, it may be necessary to plant tolerant varieties of canola. Other seedling blights may be controlled with captan, fludioxanil or thiram.

Sunflower

Downy mildew is a soil-borne disease that can cause severe losses if excessive rains occur shortly after planting. The downy mildew fungus is widespread across North Dakota. It survives many years in the soil and infects emerging or recently emerged seedlings when the soil is saturated. Several new races have occurred in recent years and at present only one hybrid is resistant to all races.

The downy mildew fungus has developed resistance to mefenoxam, metalaxyl and oxadixyl. Seed treatment with these products provides poor to fair control of downy mildew, depending on what per cent of the downy mildew population is resistant. No suitable replacement fungicide was available for the 2000 growing season.

BIOLOGICAL CONTROL

Kodiak

Kodiak concentrate contains *Bacillus subtilis* bacteria which colonize the developing root system, suppressing disease organisms such as *Fusarium, Rhizoctonia, Alternaria* and *Aspergillus* that attack root systems. When used with a chemical seed treatment, the combination of chemicals and Kodiak provides protection to the root for a much longer time than with chemicals alone.

As the root system develops, the bacteria grow with the roots extending the protection throughout the growing season. As a result of this biological protection, a vigorous root system is established by the plant, which often results in more uniform stands and greater yields. Registered for seed and pod vegtables, soybeans, wheat and barley, and corn plus all other agricultural seeds.

SEED QUALITY

Quick seedling emergence and even stands are essential to maximizing the yield of all crops. The use of high-quality, disease-free seed is the first step to producing good stands.

The importance of using high-quality seed continues to increase due to changing production methods such as early planting, narrow rows, and no-till or reduced-tillage planting systems. The use of high-quality seed better ensures seedlings which emerge rapidly, tolerate adverse weather conditions, and resist disease.

SEED QUALITY AND SEED TREATMENTS

A number of factors affect the quality of seed. Seed-borne fungi, insect feeding damage, poor seed storage (temperature and humidity too high), and damaged seed coats can reduce the quality of seed. Seed quality may be reduced due to infection by fungi when harvest of the seed crop is delayed several weeks or when seed diseases occur due to wet conditions. Infected seed can be detected at harvest as discoloured, shriveled, or moldy in appearance.

Insect feeding that may occur in the field prior to harvest or in storage may affect seed quality. Germination of seed lots may be reduced by the insect partially eating the seed or opening the seed coat to fungal infections occurring at a later date. The seed coat is the primary protective layer surrounding the seed. Seed coat damage can result from rough handling during harvest, and cleaning or planting operations. Small cracks in the seed coat increase the chances of seed-rot by permitting water-soluble nutrients to escape into the soil which activate soilborne fungi that quickly enter the damaged seed.

Grain intended for seed should be handled as little as possible, dried using low temperature drying, and stored at moisture levels below 13 per cent. High humidity and temperature during storage promotes growth of fungi and insect damage. Seed lots should be closely monitored for mold growth and insect activity during storage. Seed treatments can play an important role in achieving uniform seedling emergence under certain conditions.

The selective use of seed treatments can protect seeds or seedlings from early season disease and insect pests affecting crop emergence and growth. A few seed treatments may also be used to enhance crop performance during the growing season.

Treatments containing Rhizobium inoculant have commonly been used to enhance the nitrogen fixing capability of legume crops. Increased germination can be observed when seed treatments containing fungicides are used when poor germination results from a fungal infection. Fungicide are available to protect seed and seedlings from many seed-borne or soil-borne pathogens.

Seed treatments are most beneficial for seeds infected prior to planting or when cool, wet soil conditions exist at planting which results in delayed germination. Seed treatments containing insecticides can serve to protect seed and seedlings from certain insect pests found in the soil. Cool, wet, seed beds resulting in delayed germination or plowing down green manure crops may warrant including an insecticide seed treatment.

Seed treatments should not be considered a cure-all for the selection of poor seed lots. Seed treatments will not increase poor germination due to excessive mechanical damage, poor storage conditions, genetic differences in variety, or other damage. The effect of a seed treatment on plant stands will depend upon seed quality as well as field conditions at planting. Extensive research has shown seed treatments can increase stands from poor quality seed or when planting conditions are less than optimum. However, seed treatments do not always increase yield. Generally, seed treatments result in increased yield by preventing death of seedlings causing frequent skips within crop rows.

Grain producers should realise that any one seed treatment will not control all diseases or insects that may attack seeds and seedlings. Protection is also short lived, generally only lasting as long as it takes for the crop to emerge. Before selecting a seed treatment it is essential for the grower to consider potential problems associated with the seed and the history of problems associated with each field. Each planting situation is different requiring consideration of seed quality; planting rate; tillage and seed bed preparation; soil moisture and temperature; time of planting; and the likelihood of rapid emergence.

GERMINATION TESTING AND SEED QUALITY

The purpose of laboratory testing of seed germination is to assess seed quality or viability and to predict performance of the seed and seedling in the field. Seed processed for sale must be tested by a qualified laboratory under the Association of Official Seed Analysts Rules for testing seeds. Several different kinds of testing are available depending on the type of seed to be tested, the conditions of the test, and the potential uses of the seed.

The most common tests are the warm germination test, cold germination test, accelerated aging test, and the tetrazolium test. Each test is designed to evaluate various qualities of the seed. Factors that can affect the performance of seed in germination tests include; diseased seed, old seed, mechanically damaged seed, seed stored under high moisture, and excessive heating of seed during storage or drying.

In most cases a seed treatment will improve germination of seed only if the poor quality is due to seed-borne disease. The most common test is a warm germination test because it is required by seed laws to appear on the label. The percentage of germinating seed in a warm germination test must be printed on the label of the seed if it is to be sold as seed. Germination is defined as: "the emergence and development from the seed embryo of those essential structures which are indicative of the ability to produce a normal plant under favorable conditions."

The warm germination test reflects the stand producing potential of a seed lot under ideal planting conditions. Usually 400 seed from each seed lot are placed under moist conditions on blotters, rolled towels, or sand and maintained at 77 degrees F for about seven days. At the end of this period the seedlings are categorized as normal, abnormal, or diseased, and dead or hard seeds. The percentage germination is calculated from the number of normal seedlings from the total number of seeds evaluated.

The cold germination test is designed to measure the ability of seeds to germinate under high soil-moisture content and low soil temperature. This vigour test simulates early season adverse field conditions and usually represents the lowest germination that would be expected from a seed lot planted under such conditions. Actual field germination would normally fall between the cold test result and that of the warm germination test. Seeds are planted in a sand-soil mix at high moisture content and maintained at 50 degrees F for seven days. The test is then placed at 77 degrees F for four days. The percentage of healthy seedlings that emerge at least one inch above the soil is reported.

The accelerated aging test estimates the carryover potential of a seed lot in warehouse storage. The seeds are exposed to high temperatures and high relative humidity for short periods of time that cause seed deterioration. Seed are suspended over water in a chamber for 72 hours (wheat and soybeans) or 96 hours (corn) then tested in a standard warm germination test.

This test only would be used on seed whose longevity was in question. The tetrazolium test is a "quick test" for seed lot viability. It is useful when an approximate germination percentage is needed immediately. Seeds are soaked overnight in water then treated with tetrazolium to give an indication of viable, abnormal, and dead seeds in the seed lot. This test will not detect seed-borne disease, thus is limited in its ability to estimate seed quality. This test is highly reliable for determining viable seed of corn, wheat, oats, barley, and other grasses.

If growers wish to use bin-run seed and has not had a germination test conducted by a competent laboratory, they can get an approximate germination test using the following procedure in their own home. Place two paper towels in the bottom of a dish or tray, one on top of the other; wet the towels thoroughly and tilt the tray up on one end so that excess water runs off the tray. Select a random sample of 100 seeds from the seed lot and place them in between the moist paper towels. Put the tray in a plastic bag and tie the end shut to prevent the towels from drying out.

Place the tray in a location of diffuse, not direct light, such as a north window. The location should be warm enough for good plant growth. An ideal location is with well growing house plants. After five to seven days, open the plastic bag and count the number of germinated seed with intact tap roots and shoots. Do not count moldy seed or diseased seedlings. Testing 400 seeds in this way will give a good indication of the germination percentage.

Choose the Right Fungicide for the Problem

The basic component of most seed-treatment products is a fungicide. Insecticides are contained in some products to give insect protection when such control is desirable. The grower should select a seed treatment product that will control the most likely pest to be encountered within a given field. It is important to note that not all fungi are alike.

One fungicide seed treatment may be very effective against one fungus but will have no effect on another. Rates of active ingredient used also affect a seed treatment performance on different disease problems. Fungicides have fungicidal activity, that actually kills the fungus, or fungistatic activity, that slows the growth of the fungus to prevent extensive damage. Fungicides used in seed treatment are of three types: protectants that generally prevent infection of seeds and seedlings by soilborne fungi, surface disinfestants that kill fungi and fungal spores on the surface of the seed coat (seed-borne fungi) and "systemic" disinfectants that kill fungi already established within the seed (seed-borne fungi).

Insecticides used in seed treatments act as surface protectants killing insects as they feed on the seed or emerging seedling. The marketing of seed treatments by manufacturers may result in the same chemical being sold under several brand names and by several different companies. All chemicals have three names on the packaging; chemical name, common name, and trade or brand name. The chemical name represents the molecular structure of the active ingredient and is often long and complex.

The common name is used by the manufacturer in lieu of using the longer chemical name. The trade or brand name is used by the manufacturer in product marketing. The chemical name (R,S)-2-[(2,6-dimethlyphenly)-methoxyacetylamino]-proprionic acid methyl ester has the common name mefenoxam and is marketed under the trade or brand name Apron XL LS. Using mixtures of different seed treatment fungicides or insecticides is a common practice. Mixing two or more fungicides or insecticides results in a product that will control a broader spectrum of seed pests.

Applying Seed Treatments

Seed treatments come in a variety of formulations:

- Dry flowables (DF),
- Liquid flowables (LF),
- True liquids (TL),
- Emulsifiable concentrates (EC),
- Dusts (D),
- Wettable powders (WP).

These materials may be used in slurry and mist type seed treaters upon mixing with water as specified in instructions on the product label. Special adhesives are incorporated into the dust formulations to adhere pesticide particles to the seed surface.

Commercial Treatment

Commercial seed treatment is often desirable due to the specialized equipment required to properly apply treatments or to treat large volumes of seed. The primary concern of the commercial treater is equipment calibration to ensure delivering the proper amount of active ingredient on the seed. This has become especially important with more modern fungicides that require only very small amounts of material (less than 0.5 oz) per hundred weight of seed. Detailed information on the process of treater equipment calibration can be obtained from the equipment manufacturer. Seed treatment manufacturer labels provide information related to mixing formulations and mixing procedures. Information needed to properly calibrate a seed treater includes:

- The correct labeled rate of the seed treatment material that will be used,
- How many pounds of seed is dumped each time the weigh pan arm dumps,
- The size of the cup being used in the treater.

This information will allow you to calculate the correct application rate for the seed and treatment material used. A new generation of seed treater is becoming available that offers precision distribution of treatment chemicals on individual seed through atomization. These advanced controlled droplet application (CDA) machines offer complete seed coverage with lower water volume, shorter drying time, and continuous seed flow. These machines also reduce chemical exposure to employees and are capable of applying current and future products that may include both chemicals and biologicals.

On-farm Application

Many seed treatment materials are available for on-farm use. These are known as hopper-box or planter-box treatments. Most hopper-box treatments are dry treatments formulated with talc or graphite which adheres the treatment chemical to the seed. Liquid hopper-box treatments are available as a fast-drying formulation.

Good seed coverage is required for maximum benefit from any seed treatment formulations. Obtaining thorough seed coverage can be difficult when attempting to treat seed in the field at planting time.

Most seed treatment companies recommend that the planter box be filled only half full with seed, and half the required amount of treatment product added. Mix the product into the seed volume with a stick. Then add the remainder of seed to the planter box and add the rest of the required amount of product, again mixing thoroughly.

Good results are also obtained by premixing seed with the product in a suitable container, then pouring treated seed into the planter box. On-the-farm treaters are available that apply liquid or dry formulations to seed as it passes through an auger from the transport bin or truck to the planter boxes. These treaters are a very convenient way to apply seed treatment onto bulk seed. For additional information on on-the-farm treaters, contact a representative of the company selling planter-box treatments.

Labeling Treated Seed

Once seed is treated it can no longer be used for food feed or other purposes. Under the federal and state seed laws, the following information is required to be shown on the label: a word or statement in type no smaller than eight points indicating that the seeds have been treated; the commonly accepted, chemical (generic) name of the applied substance including the application rate, and a caution statement if the substance used in such a treatment has an amount remaining with the seed that is harmful to humans or other vertebrate animals.

Seed treated with highly-toxic substances shall be labeled in red ink to show a statement such as "Poison Treated." In addition, the label should show a representation of skull and crossbones. Seed treated with less toxic substances shall be labeled to show a caution statement in no smaller than eight point type such as "Do not use for food, feed, or oil."

Colouring Seed Treated with Poisonous Substances

All treated seed must be coloured an unnatural colour to distinguish it from untreated grain and prevent inadvertent use as food for man, feed for animals, or for oil purposes. All treated seed must be coloured with a dye, colour coat, or colour film which must cause no apparent injury to seed germination or danger to personnel processing or using the seed. Dyes are designed to stain the seed with colour.

Colour coat pigments cover the seed with colour, and colour films are made from a polymer that creates a coloured film around the seed. Seed treatment companies currently offer a wide variety of colour and surface texture options to seed processors.

DISEASES OF ASPARAGUS

Three diseases caused by fungi affect asparagus in Ontario. Fusarium root and crown rot is caused by a soil-borne fungus, while asparagus rust and Botrytis blight are caused by fungi whose spores are primarily carried about by wind. The life cycles and characteristics of these fungi will be discussed in this Factsheet to develop an understanding of how they may be recognized and controlled.

FUSARIUM ROOT AND CROWN ROT

Description

The major disease problem of asparagus is caused by two species of fungi called *Fusarium*. *Fusariummoniliforme* causes decay of storage roots, stems and crowns. This fungal species is present in all agricultural soils and infects corn, grasses, and other monocotyledonous plants as well as asparagus. *Fusarium oxysporum* f. sp. *asparagi* causes root rot and seedling blight, and may also plug the water conducting vessels, causing wilting of spears and fern.

These species of soil-dwelling fungi are very prolific, long-lived, and capable of growing as saprophytes, feeding on decaying asparagus residues and soil organic matter.

They colonize old roots and crowns, invading directly through root tips or through wounds from implements, cutting tools, or insect feeding. Asparagus plants which are under stress are more susceptible to infection than those which are growing vigorously.

Affected spears may shrivel and rot in spring before or after emergence. Infected crowns have hollow, rotted feeder and storage roots. When crown and stem tissue is sliced open, a reddish-brown discoloration is visible. Symptoms on fern includes stunting, yellow to brown discoloration of one or more stalks per crown, and fewer stalks per crown. Affected crowns decline in vigour and die. Such damage is scattered throughout the field and increases until the stand is too sparse to harvest economically.

Fig. Yellow Discoloration of Asparagus Fern Infected by *Fusarium*.

Fig. Cross-section of Asparagus Crown Infected by *Fusarium*.

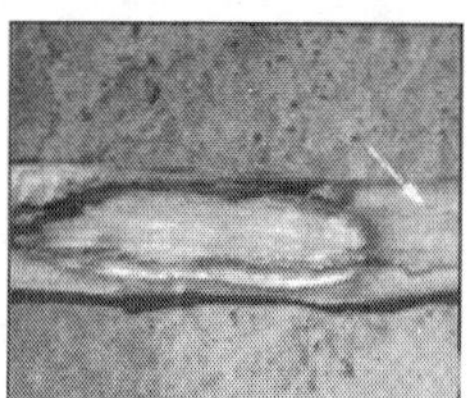

Fig. Cross-section of Asparagus Spear Infected by *Fusarium*.

CONTROL

Because *Fusarium* species are present in most agricultural soils, the diseases are almost impossible to avoid. Control is therefore directed at minimizing infection early in the life of crowns, and at maintaining a vigorous, long-lived asparagus stand by careful management.

Suggested management practices include:

1. Avoid replanting in land which previously grew asparagus. *Fusarium* builds up to extremely high levels during the long period of asparagus culture in the field, and survives many years in that soil even after the crop is removed.
2. Soil fumigation with a suitable material will reduce infection in crown nurseries or direct-seeded asparagus.
3. Use treated seed. Untreated seed should be disinfested with 1 part sodium hypochlorite bleach in 5 parts water for two hours. Rinse seed in fresh water and dry before planting.

4. Use only vigorous, one year old crowns and encourage proper handling procedures for transplanting. Weak crowns are highly susceptible to infection.
5. Minimize stress in young and established plantings. Stresses which weaken asparagus crowns and thus promote *Fusarium* diseases are:
   - Overextending the harvest season, or picking before crowns are well established;
   - Insect and disease damage;
   - Weed competition;
   - Mechanical wounding from tillage equipment, cutting knives, etc.;
   - Poor soil drainage;
   - Acidic soil pH;
   - Drought;
   - Injury from misapplication of pesticides, fertilizers, and soil ammendments;
   - Low fertility; and
   - Soil compaction.
6. Use resistant varieties. At present, no varieties recommended in Ontario are resistant. However, lines and varieties being developed in the University of Guelph breeding programme are being screened for *Fusarium* resistance.

ASPARAGUS RUST

Description and Life Cycle

Puccinia asparagi, which causes asparagus rust, has a complicated life cycle consisting of several stages, all of which occur on asparagus. Some members of the onion family, such as cooking onions and chives, are also susceptible. There are no alternate hosts such as is common with other *Puccinia* rusts on wheat and oats.

The fungus overwinters as teliospores on asparagus debris. Teliospores germinate in spring, producing small, basidiospores which are blown onto emerging spears and cause infection. Later, from April until July, small upraised, light-green, oval lesions (patches) called aecia occur on the lower portion of the infected fern stalks.

As the aecial lesions turn creamy orange in colour, aeciospores are released and re-infect asparagus fern during several hours of continuous leaf wetness. Twelve to fourteen days after re-infection, upraised tan blisters called uredia appear on asparagus stalks and foliage. Uredia break open to expose masses of rusty-coloured spores called uredia spores, after which the disease is named. Urediospores repeatedly re-infect asparagus from June until

September. Warm weather with heavy dew, fog, or light rainfall enhances rust development. Late in summer, telia develop, producing black teliospores, completing the yearly life cycle.

DAMAGE

The fungus develops in the tissue of asparagus fern and drains the plant of vital nutrients. The foliage then dries out and falls prematurely, further reducing the production and storage of food for the following crop. Successive years of infection by rust will weaken asparagus crowns.

CONTROL

1. Cultural control involves breaking the continuity of the rust life cycle. Inoculum from overwintered teliospores in fern residue may be spread from wild or neighbouring asparagus, uncut spears, young or seedling plantations, or volunteer seedlings in the asparagus crop. It is important to:
   - Destroy wild and volunteer asparagus;
   - Clean cut all fields during harvest; and
   - Locate nurseries and young plantings remotely from established fields if possible.
2. Use resistant asparagus varieties. Viking varieties have some degree of tolerance to rust, but will become infected under conditions of favourable environment and abundant inoculum. Breeding and variety trials at the University of Guelph are including rust resistance as an important criterion in the development of new varieties.
3. Control with fungicides is presently limited to the chemical zineb. Spraying should begin when the aecial or initial uredial stages are evident on the fern - in May on young plantings, in late June in harvested plantings. Because zineb is a protectant fungicide, adequate coverage *before* rust appears and throughout the season is essential to achieve control.

Fig. *Botrytis* Lesions on Asparagus Fern Branches. Note the Tufts of Fruiting Bodies Produced by the Fungus in these Lesions.

Fig. Extensive Blighting of the Lower fern Canopy Caused by *Botrytis*.

Fig. Aecial Lesion of Asparagus Rust. Note Oval to Elliptical Shaped, Upraised Lesion.

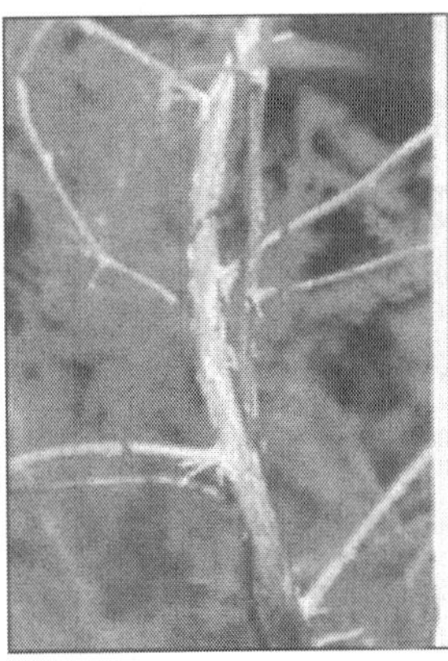

Fig. Uredial Lesions of Asparagus Rust. Note Tan Blisters which have Opened, Exposing the Rusty Coloured Spore Mass.

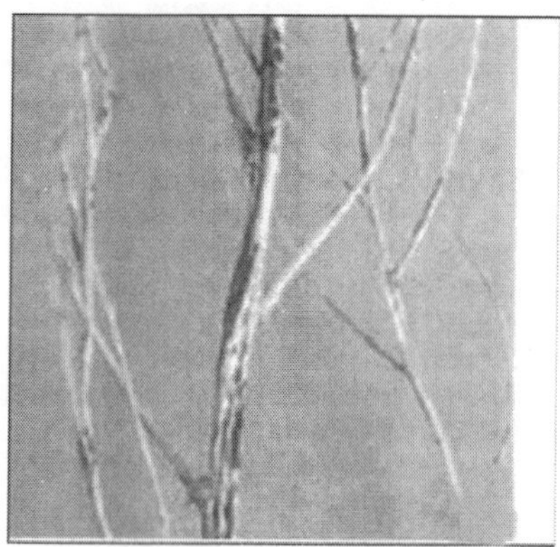

Fig. Telial Lesions of Asparagus Rust. Note Black Teliospores on Overwintering Asparagus Fern.

- Botrytis Blight
- Description and Life Cycle

This disease, caused by the fungus *Botrytis cinerea*, occurs during summer, causing browning of the lower fern canopy (Figure). *Botrytis* progresses most rapidly during hot, moist weather when the fern does not dry adequately. This fungus attacks many other crops, such as strawberries and other small fruit, potatoes, beans, and other vegetables, and many ornamentals.

The disease begins on senescing (dying) flowers or injured fern. Botrytis spores are spread by wind and rain within the dense fern canopy. Individual lesions are tan with dark brown borders, often surrounded by a fellow halo (Figure). When wet weather persists, newly-emerged spears may be completely blighted, turning brown or black, and often covered with grey, fuzzy, spore-bearing fungal growth.

Control

Zineb will also give some control of *Botrytis* when used regularly for rust control.

Viruses

Though several viruses have been identified in asparagus in Europe and Western USA, they have not yet been reported in Ontario.

Pesticide Application

As fern growth becomes dense, more water and higher rates of pesticides must be used to obtain adequate spray coverage. Good coverage is essential, especially for protective fungicides like zineb. Drop nozzles are recommended. Up-to-date pesticide recommendations are available for commercial growers in OMAFRA Publication 363, *Vegetable Production Recommendations*. Homeowners are referred to OMAFRA Publication 64, *Insect and Disease*

Control in the Home Garden. In conclusion, as a perennial crop, asparagus is a long-term investment. It will pay to monitor pests regularly throughout the harvest and growing seasons. Only then can problems be detected and controlled before they cause long-term economic damage.

SAFETY PRECAUTIONS

Seed treatment chemicals are pesticides and should be handled as such. Product labels provide information on safe handling and application. Be sure to follow all instructions as listed.

Hazards associated

Read the label and follow instructions carefully. Over treatment may injure the seed and under treatment will not control the pests in question. The label will also provide information on the toxicity of the active ingredients in the form of code words. The label will provide basic information on first aid, environmental hazards, directions for use, proper storage and disposal of containers, and pesticide and general product information.

Use the label and be familiar with it. Material safety data sheets (MSDS) are available from the company for each of the products sold. Consult the MSDS for more detailed information before applying the material. When handling toxic substances, operating personnel must be provided with protective clothing such as coveralls, cap, protective glasses, rubber apron, rubber boots, rubber gloves, and a respirator designed for use with the substance. Personnel should not inhale the dust or vapour nor permit the material to contact the skin or eyes.

The operator should wash thoroughly with soap and water before eating and smoking. Bathe immediately after work and change all clothing and wash clothing thoroughly with soap and hot water before reuse. In case of contact, immediately remove contaminated clothing and wash skin thoroughly with soap and water. A safety shower should be installed in the immediate vicinity of the treater. An exhaust system should be installed to remove vapours and dust from the operating area. The exhaust air should discharge into a cyclone or bag-type dust collector. The treater should also be vented and tied into the exhaust system. Seed treaters should be isolated and not operated in the vicinity of other personnel or farm commodities that are to be used for food, feed, or oil.

Special multiwall paper bags or tightly woven bags are recommended for seed that has been treated with toxic substances. Seed should be thoroughly dry before going into the bag, as excessive moisture can cause rapid deterioration of the seed. Store treated seed in a cool, dry place away from food or feed products. Seed treaters should be thoroughly cleaned after use as some of the pesticides are corrosive and others settle out and cause clogging of the equipment. Do not run contaminated water into a stream or public sewer, but discharge into a shallow ground pit.

Hazards associated with the use of treated seed

The user of treated seed should read and follow the label carefully. The planter hoppers should be filled outdoors. Do not breath dust or fumes from treated seed or get in eyes or on skin. Wash thoroughly with soap and water after handling treated seed. Use treated seed as seed and do not introduce them into the food or feed channels because serious injury to livestock and humans may occur. Follow instructions on the label for disposal of seed treatment containers and bags containing treated seed.

Do not reuse pesticide containers. Triple rinse and, if possible, recycle containers. Containers can be punctured and disposed of in a sanitary landfill or incineration, or, if allowed by state and local authorities, by burning. If burning, stay out of the smoke. Old seed with low germination should be destroyed by burying at least 18 inches deep in an isolated area away from the water supplies. Treated seed exposed on soil surfaces will be hazardous to birds and other wildlife.

Biological Seed Treatments

Recent research has focused on biological control of certain seed and seedling diseases of crops. The most promising biological control agents tested have been either fungi or bacteria. These biological seed treatments control seed pests by parasitizing the pest organisms, competing for food on the root system, or producing toxic compounds that inhibit pathogen growth. Several biological seed treatments are currently on the market.

Insecticide Seed Treatments

A limited number of insecticides are labeled for use as seed treatments to protect seedlings of various field crops from soil inhabiting insects.

SOIL INSECT PESTS

Soil insect pests are generally more of a problem on corn and soybeans than on small grains and forages. Soil-applied insecticides recommended for control of rootworm larvae also may prevent early stand losses due to wireworms, grubs, seedcorn maggot, and seedcorn beetles. If a soil-applied insecticide is used at planting, then a seed applied insecticide may not be necessary.

Refer to the label of the soil-applied insecticide to determine which insects are controlled before purchasing a seed treatment insecticide. Stand losses to these soil insects is generally not significant enough to warrant routine use of soil-applied insecticides in first-year corn. The use of insecticide for first-year corn may change in the future due to a biotype of the western corn rootworm beetle. Benefits from suppression of early-season soil pests may be significant in:

- No-till or reduced-tillage conditions with poor weed control where

a high risk exists for a given insect problem such as wireworms and grubs in corn following established sod,
- Continuous corn where early-season pests may cause unacceptable stand losses.

Seedcorn Maggot

Seedcorn maggot or the bean seed maggot are small, yellowish-white, legless fly larvae found feeding on corn seeds. Extensive feeding by these maggots will cause a reduction in stand. In general, seedcorn maggot problems are most likely to occur in situations where:

- High organic matter or decaying vegetation attracts egg-laying adult flies.
- Cool, damp soil conditions delay seed germination and prolong the period vulnerable to maggot attack.

No-till production systems have not been found to increase the likelihood of problems with seed maggots. Several products are formulated for application prior to planting seed on the farm as planter-box or hopper-box treatments.

Seedcorn Beetle

Fig. Seedcorn Beetle

Partially eaten seeds, loss of germination, or stunted seedlings in the presence of small (1/4 to 1/3 inch long) brown ground beetles indicate a seedcorn beetle problem. As with seedcorn maggots, damage is most likely to occur under cool, damp conditions where seed germination and seedling development are delayed.

Wireworms

Wireworms are the larval stage of the "click beetles." The term "wireworm" applies to a complex of species with life cycles requiring one or more years per generation. Wireworm populations affecting corn are most severe in fields following sod or fields having a prolonged grassy weed problem. The larvae pass through a number of life stages, or instars. The earliest stages are very small and white; the latter stages have a characteristic hard-shell appearance and are yellow-brown in colour.

Full-grown larvae range from 1/2 to 1 inch long depending on the species. Wireworm's damage corn by feeding on germinating seeds and young seedlings and may bore into stalks at the soil level. Stand loss may be significant in fields with high populations. In fields where a soil-applied insecticide is not used, especially first year corn fields, application of seed treatment insecticide is strongly recommended.

WHEN TO TREAT SEED

Corn

Treatment is recommended for all seed corn to prevent or reduce seed decay and seedling blights. The germination and early growth of the seedling is a critical period in the life cycle of corn. Soilborne organisms may invade and kill the embryo before germination, or when invasion occurs later, the seedlings may be destroyed before or after emergence. Seedlings surviving fungal infections at germination are often less vigorous.

Corn seedling diseases are more prevalent in cold, wet soil than when soil temperatures are above 55 degrees F. Therefore, early planted corn needs the added protection of a good seed treatment fungicide. Seedling blights tend to be more severe in fields with poor seed bed preparation such as in no-tillage or reduced-tillage fields. Corn seed is generally treated with a fungicide or fungicide-insecticide combination by the seed producer or seed processor. Corn smut, leaf blights, stalk and ear rots, and virus diseases are not controlled by seed treatment fungicides.

Small Grains

Seed treatment of small grains is recommended to control smut diseases and to reduce seed decay and seedling blights. The smut diseases include stinking smut (common bunt) of wheat, loose smut of wheat and spelt, and the loose and covered smuts of oats and barley.

All of the wheat varieties planted are susceptible to stinking smut (common bunt). Wheat and oat varieties have resistance to some, but not all, races of loose smut (wheat) and loose and covered smuts (oats). The only sure method of smut control is by using an effective seed treatment fungicide. Spelt is extremely susceptible to loose smut and all seed should be treated with an effective fungicide.

Seedling blight phases of head scab (*Gibberella zeae*) and glume blotch (*Stagonospora nodorum*) contribute to poor stands of small grains. Both seed infecting fungi contribute to poor quality seed by causing lightweight, shriveled grain. Since these seed-borne diseases are common, all small grain seeds should be treated to control seedling blight when infected seed are planted.

Soybean

Seed treatment for soybean is recommended in three disease situations:

- When poor quality seed is used for planting,
- When early season control of Phytophthora damping-off is needed,
- When planting early into cool, wet soils, especially in reduced tillage.

The major cause of poor-quality soybean seed is Phomopsis seed rot. Infected seed may be visibly moldy, yet others appear healthy. Purchasing certified seed with the per cent germination, the variety name, and the seed treatment material used listed on the label is one way to ensure quality seed is planted. Germination percentage of bin-run beans is generally unknown unless they are tested.

Seed lots with greater than 80 per cent germination generally do not benefit from seed treatment when planted at recommended seeding rates in warm soil. Those with less than 80 per cent germination may benefit from seed treatments by increased stand and, possibly, increased yield. Decisions to use poor-quality seed should be based on the fact that any seed treatment fungicide effective against seed-borne diseases will not increase germination more than 20 per cent. If bin-run beans of 50 per cent germination are treated with a fungicide, the grower should not expect over 70 per cent germination in the field. Planting rates can be increased to compensate for seed of low germination.

A number of fungicides are effective in controlling Phomopsis seed rot. These materials have varying degrees of activity against seed and seedling diseases. Phytophthora damping-off is a serious disease of soybean seedlings in the more heavy, poorly-drained soils. The increase in no-tillage and reduced tillage has increased the incidence of Phytophthora damping off in the state. No-till soils remain wetter longer in the spring and less precipitation is required to saturate them compared with plowed soil during the early stages of crop development.

Extra moisture in the soil pore spaces favors germination of *Phytophthora*. It is recommended that seed treatments be used even when planting *Phytophthora* resistant varieties to ensure good stand establishment. Rhizoctonia seedling blight has caused considerable stand losses over the last decade. Damage from *Rhizoctonia* appears to be a stress related in that factors that adversely affect germination and establishment of seedlings predisposes the plants to infection.

Stress from planting in dry soil, herbicide injury reduced tillage fields and shallow planting appears to increase the potential for damage by *Rhizoctonia*. Adequate seed bed preparation, soil fertility, and soil moisture for rapid emergence lessens the potential for problems from *Rhizoctonia*.

Sclerotinia stem rot may be introduced into new fields by sclerotia-contaminated seed lots or infested seed. Seed should be well cleaned to remove sclerotia and soybean seed should be treated with a fungicide to eradicate the fungus on infested seed.

Alfalfa and other Small-seeded Forage Legumes

Although a number of fungicides are registered for use as seed treatments of alfalfa, clover, etc., these crops usually do not respond to seed treatments under field conditions. Some increases in stand may result but these increases have not been reflected in increased yield.

The one notable exception to this is the use of fungicide seed treatment on alfalfa for control of seedling damping-off caused by *Phytophthora*. Significant improvement in stand establishment of alfalfa has been noted in research plots using treated seed. In areas where Phytophthora root rot has caused a serious problem in reducing alfalfa stands or preventing establishment of alfalfa, a variety with resistance to *Phytophthora* that has been treated with a *Phytophthora* specific seed treatment is highly recommended.

Rhizobium Inoculation and Seed Treatments

Rhizobium and Bradyrhizobium are bacteria which can fix nitrogen from the air or collect nitrogen from the soil water solution and incorporate it into compounds that plants can use.

The nodules that form on soybean roots are the sites where the bacteria reside. Rhizobia are present in soils, but the commercial strains and some newer strains identified by the USDA are much more efficient at fixing nitrogen. The bacteria may be applied to the seed or placed in the seed furrow at planting. Insecticide and fungicide seed treatments can have a negative impact on inoculants applied to soybean seed.

Several factors affect the impact of seed treatments on seed applied inoculants including: toxicity of material, toxicity of carrier or formulation, length of time inoculant is in contact with seed treatment, and formulation of inoculant. Insecticide seed treatments tend to be the most toxic of all seed treatments to inoculants. The fungicides themselves are not necessarily the problem, but the formulation or carriers may inhibit Rhizobial growth and colonization.

Using fungicides in combination with inoculants can be successful if some precautions are taken. If a specific material is highly toxic to inoculants, then placing the inoculant directly in the furrow and not on the seed is recommended. Treating the seed first and allowing the seed treatment material to dry followed by using humus preparation of inoculant may be successful. Liquid inoculants and fungicides should not be mixed and applied simultaneously. Producers should minimize the time the inoculants are in contact with the seed treatment fungicides prior to planting.

Corn Diseases

Damping-off and seed decay. Germinating corn seed can be attacked by several seed-borne and/or soilborne fungi. Pre-emergence and post-emergence damping-off caused by fungi is most common in poorly drained, cold soils. Seed rots and seedling blights are commonly caused by *Pythium* and *Fusarium* species, and may be caused by *Penicillium* and *Bipolaris* species. All of these fungi can rot seed prior to germination. Infected seedlings typically show a marked softening of stem tissue at the soil line.

Kernels with surface cracks caused by mechanical harvesting are especially susceptible to seed rots caused by soilborne pathogens. Broad spectrum, protectant-type fungicides are highly recommended for both corn and sorghum, especially when early planting in cold, wet soils is attempted. Pythium seedling blight is most significant under these conditions. Most seed corn companies treat their seed prior to bagging. However, planter-box formulations of these materials are available for use by the grower on untreated field, sweet, and popcorn seeds.

Soybean Diseases

Phytophthora damping-off and root rot is caused by the fungus *Phytophthora sojae*, and is particularly a problem in low, poorly drained, clay soils. However, the disease can be encountered on a variety of soil types if the soil remains wet for several days soon after planting. Phytophthora can attack soybean plants at any stage of development and stands can be reduced by seed rotting and pre-emergence damping-off. Young plants can be killed soon after emergence. Brown, water-soaked stems and yellow, wilted leaves are the primary symptoms of post-emergence damping-off.

Phytophthora survives in the soil as thick-walled spore, called oospores. Early in the growing season, when the soil remains wet for several days, these spores will germinate, producing a second type of spore called a sporangium. The sporangium then germinates to produce a third type of spore called a zoospore in large numbers. Zoospores are attracted to chemicals released by soybean roots and swim through a thin film of water until a root is encountered. The zoospores produce a hyphae (a threadlike structure) and infect the soybean root. The fungus grows in the roots and eventually into the plant stem. As the plant dies, oospores are formed by the fungus. Pythium damping-off can be caused by several species of *Pythium*. Pre-emergence damping-off, seed decays, and root rots of soybean all can be caused by this group of fungi. Pythium rots can occur on soybeans at any stage of plant development but are much more prevalent on seedlings. Diseases caused by *Pythium* can occur over a wide range of temperatures, but are most common during cold periods with high soil moisture.

Seedlings infected with *Pythium* often fail to emerge. Stems of infected seedlings appear water-soaked and translucent above the soil line. Diseased areas later turn brown and appear shrunken; eventually, the stem and smaller roots decay and the seedlings die. *Pythium* species survive also as oospores. During periods of high soil moisture, the resting spore germinates and infects seeds or young plants in much the same way as *Phytophthora.*

Germinating soybean seeds release a wide variety nutrients or by-products that can stimulate the growth and attract spores of the fungus. Many different seed treatment materials will effectively prevent losses from Pythium seed rot and seedling blights.

Rhizoctonia seedling blight is caused by the fungus *Rhizoctonia solani,* and may occur at any time, but a protected cool period of low soil moisture followed by warming soil temperatures with a brief rainy period favors disease development. Seedlings under stress are more susceptible to Rhizoctonia seedling blight. Example of stress are herbicide injury or desiccation during early seedling growth. Seedlings and young plants are most often affected. Infected plant stems and older roots appear reddish brown and have sunken lesions. Stems may be girdled just above the soil line; tissue thus damaged may appear cracked or cankered. In dry, windy weather, severely infected plants wilt and die rapidly.

Rhizoctonia solani survives in the soil as sclerotia. The fungal mycelium body itself may also survive in soil or in association with old plant residue. Growth of the fungus in soil depends on soil nutrient supply, pH, moisture, and temperature. As the fungus grows in soil, it can encounter and infect germinating soybean seeds or young plants. Several seed treatment materials can control the seed rot phase of this disease, however, none are highly effective against the post-emergence or stem rot phase of the disease.

Phomopsis seed rot is a common seed disease and is caused by the seed-borne fungi, *Phomopsis longicolla, Diaporthe phaseolorum* var. *sojae,* and *D. phaseolorum* var. *caulivora.* Infected seeds typically germinate poorly or not at all.

Fig. Phomopsis Seed rot of Soybean

Severely infected seeds appear shriveled, cracked, and elongated, and may be covered with a white, moldy growth. However, seeds may be infected and fail to show any symptoms at all. Infection of seeds is most common when warm, wet weather delays harvest. When moldy seeds are planted, seedling death results in poor stands.

Generally, higher soil temperatures favors development of seed rot and seedling blight. Seed treatments can increase the germination rate up to 20 per cent. Seed lots with less than 80 per cent germination should not be used.

Sclerotinia

Recent research demonstrates that Sclerotinia white mold can be introduced into new fields on infested soybean seed. In addition, this fungus, *Sclerotinia sclerotiorum*, forms hard, black, small, irregular-shaped sclerotia both inside and outside of infected soybean plants. These sclerotia are harvested with the seed or returned to the soil with plant debris at harvest. Seed should be well cleaned to remove sclerotia and treated with appropriate fungicide seed treatment to eradicate the fungus from infested seed.

Wheat and Spelt Diseases

Common bunt or stinking smut is caused by the seed-borne fungus *Tilletia laevis*. The disease causes reduced wheat yields and grain quality. Common bunt was once the most prevalent disease of wheat, however, today the disease is rarely found due to the widespread use of effective seed-treatment fungicides. Heads of diseased plants may appear stunted and are noticeably thinner and darker in colour than healthy, disease-free heads. "Bunt balls" (kernels containing millions of fungal spores) replace the kernels in the flowering heads and spread the glumes much farther apart than normal. Bunt balls approximate the size of normal kernels but tend to be more spherical and off-coloured (typically gray-coloured). When the bunt balls are ruptured at harvest, they release spores into the surrounding air and contaminate the surface of healthy seed.

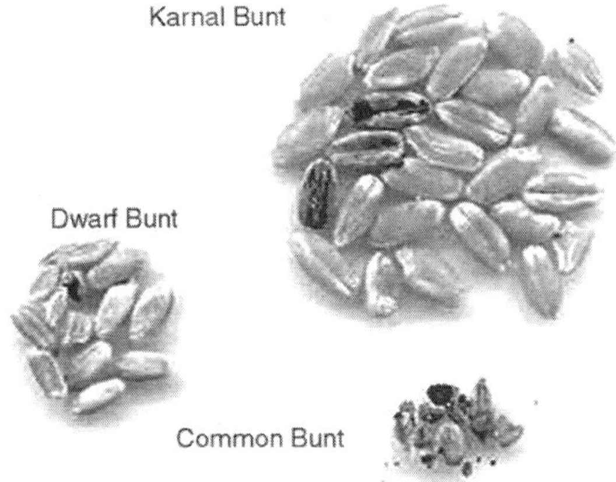

Fig. Bunt of wheat

Bunt fungi survive as resting spores on contaminated seed. When contaminated seeds are planted, bunt spores germinate in the presence of moisture and infect the wheat seedlings. Once established, the fungus grows systemically in the plant. At flowering, entire flowering heads may be infected, and the fungus converts the developing kernels into bunt balls. Healthy seeds are contaminated by the release of the spores at harvest. Loose smut, caused by the fungus *Ustilago tritici,* has little effect on seed quality. However, it can result in substantial yield losses if left unmanaged. Symptoms are most noticeable between heading and maturity of the crop. Diseased flowering heads are conspicuously darker when compared to healthy, green heads. Also, diseased heads typically emerge earlier. Diseased spikelets may be entirely transformed into dark fungal spore masses.

As the head emerges, floral tissue is torn and the spores are released. Eventually only a spike is left where a normally healthy head should be. Spores of the fungus are dispersed by wind to nearby healthy flowers where they initiate new infections during wet weather. Unlike bunt, the loose smut fungus infects the developing kernels without causing noticeable damage. *Ustilago tritici* survives in infected wheat seed.

The fungus becomes active when the seed germinates and grows into the growing point of the developing wheat plant. Fungal growth closely follows the plants' growth. When the flowers form, the fungus again sporulates and starts the cycle again. Use a systemic fungicide seed treatment to eliminate this pathogen from infected seed.

Head scab, caused by *Fusarium* spp., is frequently a major cause of poor quality wheat seed. Infected seed appear whitish to reddish in colour and are nearly always shrunken and light in weight. Scabby kernels may be dead or the young seedling may be killed by seedling blight after emerging from the soil. Roots of plants killed by seedling blight appear light to reddish brown in colour and may be covered with mold. If they survive, they generally lack vigour and may produce only a few weak tillers.

Planting infected seed early when soil temperatures are above 60 degrees F generally increases loss from seedling blight. The best control of the seed-borne phase of scab is to first clean seed lots to remove all lightweight seed, thus increasing test weights; and second, treat with a fungicide that is effective in controlling seedling blights.

Stagonospora glume blotch, caused by the fungus *Stagonospora nodorum* has increased in the US in recent years probably due to expanded use of nitrogen fertilizers, semidwarf wheat varieties, and reduced tillage. Symptoms can develop throughout the growing season on all above-ground plant parts. Initial symptoms are small chlorotic flecks usually on lower leaves or those in contact with soil. The glumes of the wheat heads become infected in late spring before flowering. Lesions generally begin at the glume tips and are dark brown in colour.

Fig. Head Scab of Wheat

During favorable weather, the fungus penetrates the glumes and infects the seed causing severe shriveling of the grain. Stagonospora survives on seed, straw, and/or volunteer wheat. With the onset of moist weather, spores are produced and are spread to healthy wheat plants by splashing rain. Glume blotch is most common in relatively warm weather when the temperature is between 70 degrees F and 80 degrees F.

Barley Diseases

True loose smut of barley is caused by the fungus *Ustilago tritici*. The disease cycle is exactly the same as that of loose smut of wheat. Infection occurs when the flowers develop. The seeds become infected internally and must be treated with a systemic fungicide to eliminate the pathogen. Semiloose smut is caused by the fungus, *Ustilago avenae.*

The disease typically causes small losses due to the use of resistant cultivars and fungicide seed treatment. It is most common when soil temperatures are cool and the soil is dry. Distinguishing semi-loose smut from true loose smut in the field is extremely difficult. Diseased heads typically appear earlier than healthy heads. Millions of fungal spores may be formed in diseased heads.

Ustilago avenae survives as teliospores on the seed surface. At seed germination, the fungus becomes active and infects young seedlings prior to their emergence from soil. As the plants develop, the fungus grows within the growing point of the young plant. At the boot stage, the fungus transforms the kernels and glumes into dark spore masses very similar to those produced by the true loose smut fungus. Spores are readily dispersed by wind or the action of harvesting equipment. Healthy seeds are contaminated upon contact with the spores. Covered smut caused by the fungus, *Ustilago hordei*, can result in significant yield losses if left untreated. It is a relatively common disease throughout the barley-growing regions of the world. Problems with the disease occur over a wide range of temperatures and are most common under moist soil conditions.

Ustilago hordei survives as thick-walled resting spores on barley seed. Typically, spore germination coincides with seed germination, it is at this time that infection occurs. Once established, the fungus invades the actively growing tissue and keeps pace with growth of the plant. At flowering, the fungus grows through the floral tissue and forms masses of spores in place of healthy seed. Each kernel becomes a spore filled "smut ball" that is enclosed by a thin membrane until plant maturity. At harvest, spores are liberated and may contaminate healthy seed.

Barley stripe, caused by the seed-borne fungus, *Drechslera graminea,* is most common during periods of rain or high humidity. Few seeds are produced on infected plants, and losses generally coincide with disease levels in a particular field. Properly treated seed is the primary means of controlling this disease. Symptoms are first noticeable on the first two or three leaves produced by the plant and on many leaves produced later. A yellow stripe on the leaf sheath and basal portion of the leaf blade is common. Later, the stripe may extend the entire length of the leaf and the leaf may die. At heading, diseased plants are often stunted and appear a light tan colour in contrast to the larger, green healthy plants. Spikes may be deformed or may fail to emerge. At heading, spores are produced on diseased leaves and spread by wind to nearby heads. Grain produced by diseased plants is typically brown and shriveled.

Drechslera graminea survives entirely as mycelium in the outer layers of the seed. When the seed germinates in moderately moist soil at temperatures below 55 degrees F, the fungus mycelium penetrates and infects the seedling. The fungus grows within plant tissue and occupies the actively growing plant parts.

Net blotch and spot blotch, common diseases of barley, are caused by the fungi, *Drechslera teres* and *Bipolaris sorokineana,* respectively. Net blotch is most common during rainy and humid weather, and early flag leaf infections can result in both reduced grain yield and weight.

A net-like pattern on sheath leaves or flag leaves is the most recognizable symptom of the disease. Initially, small spots or streaks appear on the leaves, followed by an expansion into conspicuous longitudinal and transverse streaks, eventually combining into a network. Brown areas surrounded by greenish-yellow margins are commonly observed on infected leaves. Eventually, entire leaves may wither and die.

Drechslera teres survives in infected seed or on crop residues. Infection of seedlings is most common at 34 degrees F to 60 degrees F. Once infection is established, the fungus produces spores on leaf tissue during periods of high relative humidity. Spores are released and carried by wind to nearby barley plants where new infections can result. Infection of leaves can occur when leaves are wet for periods of 5 to 15 hours within a 46 degrees F to 91 degrees F temperature range.

Spot blotch is most commonly a problem during warm, humid weather. Yield losses of up to 36 per cent have been attributed to this disease when the flag leaf has become severely diseased early in the season and has died prematurely. Seedling blight (*B. sorokineana*) in winter barley is favored by planting early in warm soil. Delayed planting of winter barley in cooler soil helps prevent blight.

Small brown spots surrounded by a yellow halo is the characteristic symptom of the disease. Both leaves and leaf sheaths on plants of all ages can become infected. Eventually, the spots coalesce and cover large areas of the leaves. Minute fungal fruiting structures containing spores can sometimes be observed in the larger spots. Severely diseased leaves may eventually die.

Bipolaris sorokineana survives in seed or infested crop residue. Leaf infection results from airborne spores produced on the seed or on residue. Epidemics of spot blotch can occur when there are prolonged wetness periods (greater than 16 hours) and the temperature is above 70 degrees F.

Note that net blotch is more common in cooler, humid weather, as opposed to spot blotch which prefers warm, humid weather. Most common broad-spectrum seed treatment fungicides are effective in eliminating seed-borne *B. sorokineana* and *D. teres* as well as preventing losses from seedling blight caused by them.

Oat Diseases

Loose smut and covered smut, caused by *Ustilago avenae* and *U. segetum*, respectively. When disease outbreaks are severe, both yield and grain quality can be significantly reduced. However, fungicide seed treatments effectively control both diseases. Symptoms of these diseases are very similar. Smutted oat panicles emerge from sheaths as olive-brown to black fungal spore masses. Smutted panicles remain more compact than healthy ones, and usually all spikelets on a diseased plant are affected. Smutted plants are often stunted and easily detected in a recently headed field of oats.

The thin membrane enclosing the spore masses of *U. avenae* usually rupture and disintegrate after panicle emergence as opposed to those of *U. segetum*, which can persist longer.

Alfalfa Diseases

Phytophthora damping-off and root rot is caused by the fungus, *Phytophthora megasperma f.sp. medicaginis*. Standing water, poor drainage, and generally wet soil conditions favour disease development. This disease can be lethal to seedlings and established plants.

Seedlings may fail to emerge or can be killed after emergence. A conspicuous yellowing of leaves, particularly lower ones, is a characteristic symptom. Infected seedlings often wilt. Seedling tap roots have dark brown to black lesions and may be rotted 2-4 inches below the crown. A yellow discolouration of internal root tissue is often present. Tap roots may appear discoloured. The life cycle of *Phytophthora* on alfalfa is similar to that on soybean, however, the fungi are completely different and do not attack the other plant host. Seed treatments are labeled for control of Phytophthora damping off. However, to protect stands beyond the seedling stage, alfalfa varieties with resistance to the fungus should be used.

DAYLILY RUST

Daylilies (*Hemerocallis* sp.) are one of the top-selling herbaceous perennials in North America. As members of the Liliaceae family, daylilies produce showy lily-like flowers that bloom in clusters, for one day before senescing. Daylilies are native to the Old World from central Europe to China and Japan.

Most selections available to North American gardeners are cultivated hybrids of original species. Daylilies are most often propagated by crown divisions in spring or fall. These divisions are transplanted into the field or containers for further growth until sale. Daylily cultivation and breeding has become exceptionally popular, with the formation of strong national and provincial/state chapters across North America. Originating in Asia, daylily rust (*Puccinia hemerocallidis*) was first discovered within North America in the southern United States during 2000. The disease quickly spread and was found in various locations throughout the United States and in Canada during 2001. Daylily rust threatens the production and cultivation of daylilies in North America. Proper identification and understanding the biology of the disease is critical for successful disease management.

Fig. Flower of 'Pardon Me'.

SYMPTOMS

Daylily rust is caused by a fungus, *Puccinia hemerocallidis*. This fungus will only infect and colonize green, live host tissue (Figure). It can infect leaves and scapes (leafless, flower stems) of daylily plants but not roots or crowns.

Under conditions favourable for disease development, symptoms appear 3-7 days after infection. Symptoms on very susceptible varieties appear as small, yellow-orange, oval-shaped pustules. The chlorotic areas between the pustules often coalesce and become quite pronounced on more susceptible varieties severe foliar symptoms). On less susceptible varieties, pustules are not as numerous and are surrounded by tan-brown, dead leaf tissue. The pustules contain hundreds of rusty coloured summer spores (uredospores). These spores can easily be transported by wind or rubbed off onto clothing, boots or tools. It is these summer spores that cause repeat infections on neighbouring daylily leaves and scapes throughout late summer and into autumn. The fungus requires living green tissue to continue to grow and produce the rust-coloured summer spores. In the autumn, before leaves begin to senesce or die naturally, new infections will produce dark brown to black pustules that contain the resting or winter spores.

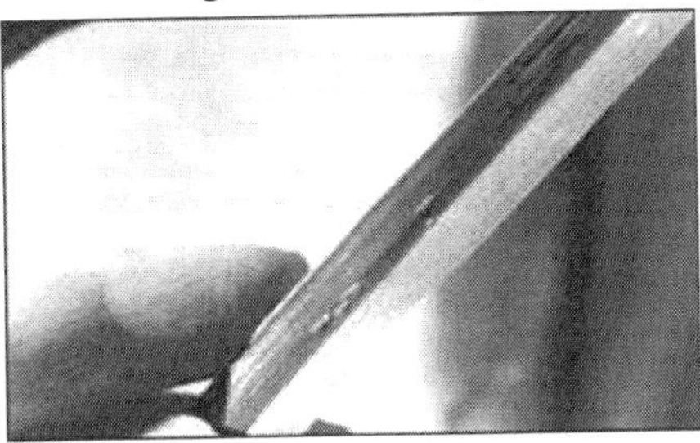

Fig. The Spores of Daylily rust can be easily rubbed off onto Fingers

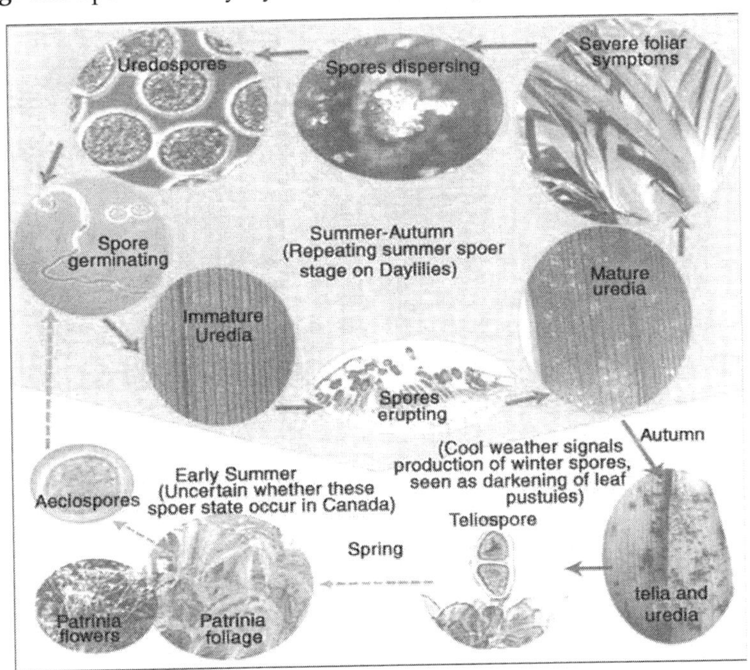

Fig. Lifecycle of Daylily Rust *(Puccinia hemerocallidis)*. The Alternate hose, *Patrinia*, is shown in the Lower Left. The Dotted Arrows Indicate the Uncertain role of *Patrinia* in the daylily rust life Cycle in Ontario

BIOLOGY AND LIFE CYCLE

The biology of the fungus *P. hemerocallidis* is complex and requires 2 different plant host species and 5 different rust spore types to complete a life cycle. In the spring, the dark winter spores (teliospores) germinate and produce another set of spores, which can only infect the alternate host,*Patrinia* spp. On *Patrinia*, 2 more spore stages are found with their own distinct symptoms (Bergeron, 2004). In summer, spores produced from *Patrinia* can infect daylilies. These infections result in yellow spots called uredia which produce the repeating summer spores (uredospores), that re-infect daylilies.

Several cycles of re-infection can occur weekly if conditions for rust infection and development are favourable. With cooler temperatures and leaf senescence, the uredia stop production of uredospores and begin to produce the dark, winter spores called teliospores. During this transition, both teliospores and uredospores can be found in the same pustules, which begin to darken as more teliospores are produced.

The masses of teliospores (called telia) overwinter, and germinate in the spring to continue the life cycle. Again, the telial stage of this fungus is likely not required for the disease to continue on into the next growing season. The fungus may be able to survive on varieties that maintain green leaf tissue during the winter, on plants brought indoors, or on plants overwintering in a minimum-heated greenhouse.

ALTERNATE HOST

The alternate host for daylily rust is the herbaceous perennial *Patrinia* spp. *Patrinia* is found in the Valerianaceae family and is sometimes referred to as Elvis-eyes or Golden Valerian.*Patrinia* spp. have small yellow flowering clusters that grow from a clump of palmately-lobed leaves, and are often used in lightly or partly shaded gardens.

The plants are native to Asia, but several species including *P. gibbosa, P. triloba, P. villosa* and *P. rupstris* are sold in the United States as well as in warmer regions of Canada. *Patrinia* is not a common perennial in North American gardens and likely poses little threat to the spread of this disease in the landscape. However, there is still some uncertainty about the role of *Patrinia* in the disease cycle of daylily rust in North America. There is very little information regarding the survival of daylily rust in northern temperate regions such as Canada. However, the disease may not require the alternate host, *Patrinia* spp., since the fungus may be able to survive on varieties which maintain green leaf tissue during the winter, on plants brought indoors, or on plants overwintering in a minimum-heated greenhouse.

Environmental Conditions for Disease Development

In Ontario, daylily rust is not usually observed until the latter part of summer and early autumn. Studies have shown that the optimal temperatures for summer spore germination can occur at 22-24°C under high humidity (although germination can occur from 7-34°C). Summer spores do not germinate in cold (<4°C) or extremely warm temperatures (>36°C) and this disease is not severe during hot, dry or cold conditions. A minimum of 5-6 hours of continuous leaf wetness is required for spore germination and leaf infection. In addition, summer spore germination decreases with high light intensity. Under conditions favourable for disease, hundreds to thousands of summer spores from each infected leaf can be produced quickly and spread rapidly, making this disease a serious threat to daylilies. This phase of the disease cycle can repeat itself many times during periods of warm weather with rain or dew periods, resulting in disease epidemics.

VARIETY SUSCEPTIBILITY

Daylily rust does not necessarily kill plants but can affect plant vigour, susceptibility to other pests and marketability. Differences in cultivar susceptibility to daylily rust have been observed in experiments conducted at the University of Georgia and the University of Guelph. However, a limited number of cultivars have been evaluated and more research into cultivar susceptibility and resistance is required. Varieties known to be very susceptible (many pustules containing numerous summer spores) include: 'Buttercup', 'Catherine Woodbury', 'Cherry Cheeks', 'Colonel Scarborough', 'Couble', 'Imperial Guard', 'Irish Ice', 'Ming Toy', 'Pardon Me', 'Karie Ann', 'Lemon Yellow', 'Little Gypsy Vagabond', 'Pandora's Box', 'Quannah' and 'Russian Rhapsody'. Moderately susceptible (fewer pustules frequently surrounded by dead brown-tan coloured leaf tissue and containing fewer summer spores) varieties include: 'Butterflake', 'Condon', 'Crystal Tide', 'Gerturde', 'Happy Returns', 'Prelude to Love', 'Joan Senior', 'Pandora's Box', 'Rosy Returns', 'Star Struck', 'Stella D'Oro', 'Summer Wine', 'Wilson's Yellow' (and*Hemerocallis fulva*). Low susceptibility (little to no infection) varieties include: 'Butterscotch Ruffles', 'Holy Spirit', 'Mac the Knife' and 'Yangtze'.

DIAGNOSING DAYLILY RUST

If you suspect your plants have been infected with daylily rust, reduce the chances of spreading this disease by placing a large, clear plastic bag over the symptomatic foliage and obtain leaf samples from within the confines of the bag. Close the bag tightly around the base of the plant and remove as much infected foliage as possible inside the bag. Take the samples indoors where foliage can be examined more closely.

Raised, orange pustules can be seen with the naked eye on both the upper and lower leaf surfaces but may be more prevalent on the lower leaf surfaces of some varieties. Pustules are more easily viewed with a 10-20x magnification hand lens. Immature pustules are elliptical, raised bumps with a waxy sheen and a yellow-orange colour.

Mature pustules have a protruding mass of orange, powdery spores. The spores can be easily rubbed off onto fingers. Submit a sample to the local Pest Diagnostic Clinic for conclusive identification.

DISEASE MANAGEMENT

- Avoid growing *Patrinia* spp. in nurseries or gardens where daylilies are grown. Spores produced on *Patrinia* spp. are thought to be infective to daylilies. If the different hosts are in close proximity, then the chances of infection are greater.
- Scout daylily plants frequently during the later weeks of summer, particularly after rain, when conditions are favourable for dew formation or during a prolonged period of cloudy days with day time temperatures around 22-24 °C.
- Select and grow the least susceptible varieties in regions where daily rust has been found.
- Avoid overhead irrigation. To minimize leaf wetness periods, direct irrigation to the soil surface and not the leaf canopy. Divide daylilies every 3-5 years and keep plants well spaced to facilitate leaf drying after rain or irrigation. If possible, irrigate plants in the morning to allow for quick drying of foliage during the day, rather than watering in the evening.
- If daylily rust is confirmed, cover infected plants with a large, clear plastic bag and remove all leaves and scapes from the infected and surrounding plants as close to the ground as possible. Composting of diseased tissue is not recommended at this time since it is not known how long the summer spores will survive under composting conditions or whether they may blow out of the compost pile and infect healthy plants nearby. Diseased tissue should be placed in a plastic bag and promptly removed from the area. Use disposable rubber gloves and wash clothing after working with or around diseased plants.
- Use registered fungicides to help protect healthy plants from rust infection. Fungicide applications should begin in early summer and may terminate when daytime temperatures do not exceed 7 °C. If plants become infected but are not yet producing powdery spore masses, systemic fungicides may be able to cure the infections and prevent pustule development.
- In the autumn, cut plants back to remove all green foliage. Do not mulch as mulching infected plants may protect rust pustules that have gone unnoticed and possibly allow them to survive the winter.

8

Field Management in Seed Production

Careful field management in seed production can significantly mitigate seed-borne diseases in seed crops. Many of the same practices that prevent disease in field crops are used to prevent disease in seed crops. Disease promoting conditions are avoided by not using overhead irrigation under moist conditions, using drip irrigation, and cutting water to allow seed crops to dry before harvesting.

Spacing plants adequately and orienting rows with the prevailing winds aids in increasing air flow—an important factor in minimizing disease. Rotating crops is important to avoid disease build-up in the soil. Of course, starting with clean, disease-free seed for planting stock is also crucial. Additionally, seed growers must use care in harvesting and cleaning seed to avoid post-harvest infection.

Seeds of Change works with our seed growers to select varieties for disease resistance. In this programme disease is intentionally allowed to build up in an isolated field for breeding and selection purposes. Seed crops are selected under the disease conditions for resistance to the disease. In this case care is taken to ensure the stock seed is not harboring seed-borne diseases when used as planting stock for seed production.

Many of Seeds of Change beet varieties have been selected in this manner for resistance to Rhizoctonia, a bacteria that causes scarring and lesions on beet roots. Many of the lettuce varieties offered have also been selected for resistance to Downy Mildew and Sclerotinia. Seeds of Change growers regularly practice rouging (selection) in seed production fields by removing diseased plants from the field.

Over 1,500 microorganisms have been shown to be related to seeds of all types. However only a small percentage of these organisms pose a threat of disease to the following crop. The majority of common diseases are not seed-borne, but initiate from the local environment.

As most diseases are regionally based it is valuable to check with local extension services and growers to identify which seed-borne diseases pose a threat in your ecosystem and then place extra care in sourcing varieties of these crops. If there are crops of particular concern, ask your seed company for recommendations and information about their disease management programme. And always use good organic disease management practices in the field.

CAUSE

Black rot is caused by a bacteria, *Xanthomonas campestris* pv. *campestris,* that can infect most crucifer crops at any growth stage. This disease is difficult for growers to manage and is considered the most serious disease of crucifer crops worldwide.

The disease can cause significant yield losses when warm, humid conditions follow periods of rainy weather during early crop development. Late infections can provide a wound for other rot organisms to enter and cause significant damage during storage.

Fig. Black rot Infected Cabbage.

SYMPTOMS

Symptoms of black rot vary considerably depending on the host, cultivar, plant age and environmental conditions. The bacteria can enter plants through natural openings and wounds caused by mechanical injury on roots and leaves. Seedborne bacteria infect the emerging seedlings through pores on the margin of the cotyledons and then spread systemically through the seedling.

Infected seedlings grown in the greenhouse under cool conditions (below 15–18 °C) frequently do not show any symptoms of the disease. When infected seedlings are transplanted to the field and temperatures rise to 25–35 °C during periods of high relative humidity (80–100 per cent), they become stunted with dead spots on the cotyledons and will eventually wilt, and die.

In regions with temperate climates (where temperatures remain cool), disease symptoms on infected seedlings may not always be obvious or appear severe. Infected seedlings grown under cool conditions may ooze bacteria from pores and lesions, which then serve as a source of the pathogen for neighbouring plants.

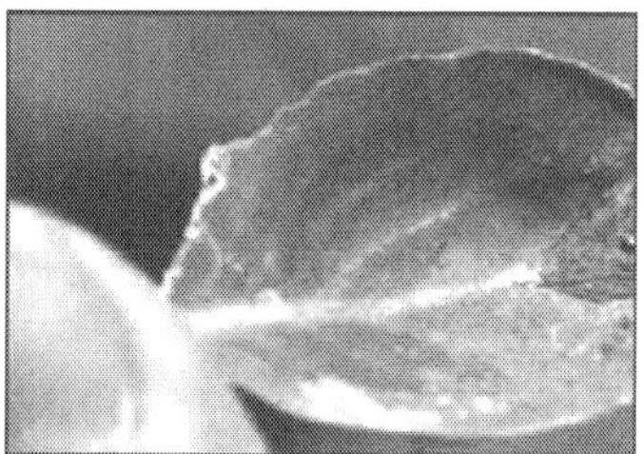

Fig. Young Cabbage leaf with V-shaped Lesion Characteristic of black rot Symptoms.

On older plants, the disease symptoms often appear as yellow or dead tissue at the edges of leaves, similar to tip burn, except the lesion frequently progress into a V-shape with the base of the V usually directed along a vein. Close inspection of infected leaves and stems may reveal black veins running through the infected tissue from which the disease gets its name. Lesions on leaves can expand down towards the base of the leaf causing the leaf to wilt and die. The bacteria produce a sticky polysaccharide called xanthan that eventually plugs the vascular tissue inside the veins causing them to collapse and turn black. The tissue above the plugged, collapsed xylem eventually turns yellow, wilts and dies. During hot humid environmental conditions, the bacteria can move from the leaf into the stem through the xylem.

Fig. Black rot Symptoms Appear as dead Tissue at the tips of (a) Kale, (b) Cauliflower, and (c) Cabbage Leaves. Note the V-Shaped Lesion Progressing from the tip along the vein of the black rot Infected Cabbage leaf.

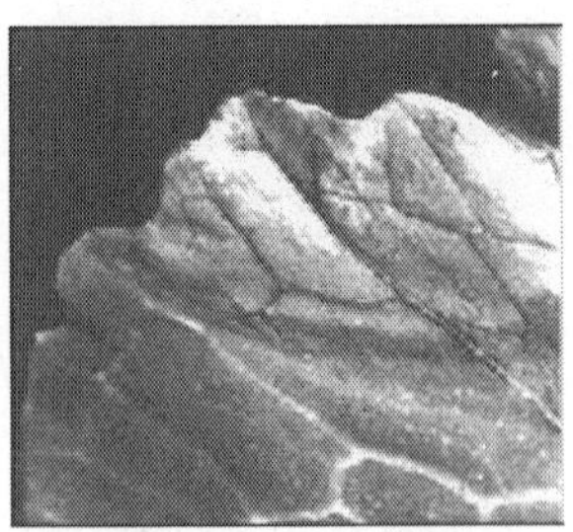

Fig. Black Rotting Veins Running through Black Rot Lesion on tip of Cauliflower Leaf.

Once inside the stem, the bacteria can move up or down to other parts of the plant including the roots. Systemically infected plants may produce chlorotic areas anywhere on the leaf. Severely infected leafy cole crops such as kale and cauliflower tend to shed their leaves from the bottom up leaving only a tuft of distorted leaves separated from the root system by a scarred barren stem. Symptoms on cauliflower often appear as black flecks or scorched leaf margins. The curds of infected cauliflower heads often become blackened. Foliar symptoms may not be visible on infected root crops such as rutabaga and radish but blackened vascular tissue can appear inside the edible root tissue rendering the plants unmarketable. Although some infected plants may appear healthy, cutting across infected stems will reveal characteristic blackened vascular tissue. This is a simple method of determining the presence of the disease.

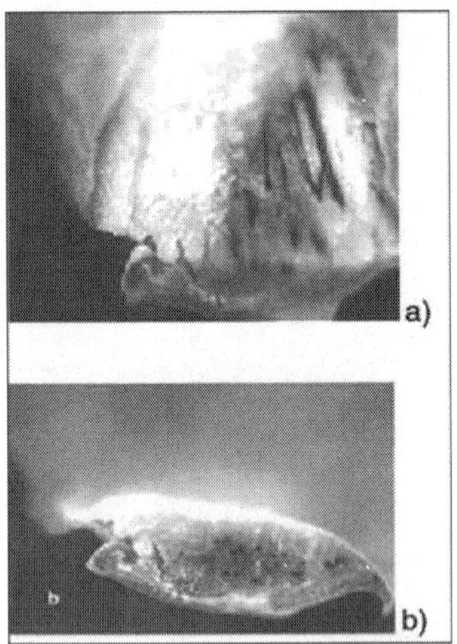

Fig. Cross Section of the base of a Black rot Infected Cabbage (a) Stem and (b) Leaf Revealing the Black, Collapsed Xylem.

Some symptoms of black rot closely resemble those caused by *Fusarium* yellows, which causes the vascular tissue to turn brown. Most commercial crucifer cultivars are resistant to *Fusarium*.

Fig. Leaf Symptoms of Fusarium Yellows Sometimes Appear Similar to black rot Except the Vascular Tissue turns brown Instead of black.

DISEASE SPREAD

Seed contaminated with black rot bacteria is considered the most important source of the pathogen and significantly contributes to the spread of this disease worldwide. As few as 3 infected seeds per 10,000 (0.03 per cent infected seeds) can result in a black rot epidemic. Seed should be tested and certified to be disease free with less than 1 in 30,000 infected seed.

The organism survives in infected crop tissue left on the soil until the crop tissue rots. However, the bacteria do not survive very long in soil as unprotected free living organisms. The black rot bacteria can also infect and survive on many crucifer weeds. This also contributes to the persistence and spread of the disease. It can grow and multiply on host tissue without infecting or causing disease.

Rain splashed bacteria from contaminated plant residue left on the soil or from neighbouring diseased plants is the primary method of disease spread throughout a field. The bacteria enter and exit through water-secreting glands called hydathodes located at the edges and tips of leaves. Hydathodes often produce a drop of water during periods of high humidity early in the morning. The pathogen spreads very quickly when rain droplets contaminated with bacteria splash onto healthy leaves and enter the hydathodes. The bacteria move into the leaf veins through hydathodes and begin to multiply, rot and plug the veins.

Fig. Hydathodes are Special Glands or Pores at the end of Vascular Tissue on leaves through which Water Exudes and are a Natural Opening for Black Rot Bacteria to Infect.

Contaminated water droplets that exude out of hydathodes of infected leaves can then be rain- splashed to other plants. Black rot is more severe and widespread in fields that receive frequent early morning rains, particularly in May and June. Equipment, people, animals and overhead irrigation can further spread the disease. Insects can also spread the bacteria; however, their contribution to the spread of black rot is limited.

DYNAMICS OF SEED MANAGEMENT

According to ENHRUM data, between 1997 and 2001, 0.5 per cent of Mexican rural farmers sowed maize seed brought from the US, but none of them conserved this seed in 2002. Nearly 3 per cent of farmers sowed maize grain obtained in Diconsa, the public retail network, at least once during the same 5-year period, but only 0.5 per cent of seed lots sown in 2002 came from this source. Seed obtained from government agencies was nearly as common as Diconsa's, while the formal seed system and other sources of grain each account for 10 times more seed. Seed exchange with other farmers through informal seed systems was overwhelmingly the main source of seed across Mexico. Its importance is much greater in the southeast than in the north, where the seed industry and other institutional sources are also significant.

RESPONSE

Farmers grow many varieties of rice. More often than not they use their own seed. Farmers have produced their own seed for generations. Farmers have exchanged seeds with neighbours for generations. This farmer seed system is also called the informal seed sector. The farmer seed system is the system that farmers use to obtain, produce, conserve, improve, and distribute seed. The farmer seed systems are important in both good years and bad, and is also important for maintaining biodiversity.

In a farmer seed system, farmers do not usually distinguish between seed and grain. Seed is not grown in separate plots but simply selected from harvested grain. Its quality can vary from very good to very bad. In the very bad, there is diseased seed, mixed seed, and weed seed. Local varieties may only be produced by the farmer seed system and so may continue to be eaten in the village due to the existence of the farmer seed system. In a village, there may be a few farmers who are recognized as good seed growers. Other farmers may receive some of their seed from these farmers. These farmers may also be looked to when a new variety is assessed. In earlier modules, you have learned how farmers can improve the quality of their own seed.

There is the formal seed sector for rice, which includes the government varietal research institutes, government seed production farms, the seed certification agency of the government, committees for varietal release, private companies, and nongovernment organizations with seed production programs and cooperatives registered by the government for seed. In the formal sector, the government develops the rules that regulate the work of each organization. An example of regulations guiding the production and distribution of seed is the Bangladesh Seed Act 1997. The farmer seed system existed way before the formal seed sector.

You will have learned under module 1 that rice is a low value, self-pollinated crop. Rice seed can be produced relatively easily and stored on-farm over a number of years. It is therefore difficult to create a stable demand for external seed supply. Because of this, private companies have not shown a high interest in producing rice seed. Therefore, the formal seed sector has tended to be dominated by the government. The farmer seed system and the formal seed sector exist alongside each other. A new variety may come through the formal seed sector and then continue to be produced each year in the farmer seed system.

RESEARCH

As calls for a 'Uniquely African Green Revolution' gain momentum, the focus on seeds and seed systems is rising up the policy agenda. Much of the debate emphasises the technological or market dimensions, with substantial investments being made in seed improvement and the development of both public and private sector delivery systems. But there is currently much less emphasis on the wider policy dimensions – and particularly the political economy of policymaking in diverse African contexts.

Experience tells us that it is these factors that often make or break even the best designed and most well intentioned intervention. And since investment in seed improvement and supply was last emphasised as a major development priority (in the 1970s and 80s), contexts have changed. The collapse of national public sector breeding systems has been dramatic, and this has only been selectively compensated for by the entry of the private sector. Large multinational seed and agricultural supply companies are increasingly dominating the global scene, and there are many claims made about the promises of new technologies (notably transgenics) transforming the seed sector through a technological revolution. While informal breeding and seed supply systems continue to exist, and indeed have been extensively supported through NGO and other projects, they are often under pressure, as drought, corruption and conflict take their toll and economic transformation and livelihood change continues apace.

The focus on *cereal* seed systems allows the research to concentrate on a similar set of crops across the four study countries with a key influence on food security at household and national levels.

Given the political reverberations of the 'food crisis' of 2007-08, this allows for a timely analysis of the implications of the policy processes shaping the breeding, production, marketing and distribution of cereal seeds. Whether grown for local subsistence or traded commercially, the significance of cereal crops to national politics (and so arguments about food security and sovereignty), commercial interests and local livelihoods – is likely to be profound.

HYPOTHESIS

The project will test the hypothesis that contrasting politics and different configurations of interests will make a difference to the way cereal seed systems operate and how a 'new green revolution' push in envisaged and ultimately plays out. The underlying implication is that politics matter and engagement with policy processes is important – defining and then deliberating among different framings and interests – *i.e.,* beyond the technical/ market fix.

A FOCUS ON SEED POLICY PROCESSES

- How do seed policies get created, by whom?
- How do ideas about what makes a 'good' seed policy evolve and change?
- How are boundaries drawn around seed problems and policy storylines elaborated?
- Whose voices and views are taken into account in the seed policy process? And what/who is excluded?
- What spaces exist for new ideas, actors, networks? How can these be opened up?

METHODOLOGY

Overall, an historical approach will be necessary to trace changes in the way policies have been framed, looking at the shifts in narratives about what the problem is and what should be done about it over time. Changes in the configuration of actors, their networks and associated interests will also help illuminate how contemporary policies have emerged. A basic mapping of the current situation will take place, involving interviews with key players (from government policymakers to public/private, national/ international researchers to commercial sector seed suppliers and traders to farmers in different parts of the country and with different resource endowments).

This approach will allow the research to elaborate (in largely qualitative terms), first, the set of 'narratives' (stories about the problems and the appropriate solutions) being deployed by different people. Second, the way such actors interact and relate will be mapped, highlighting key gaps and connections. Third, the interests of different groupings will be analysed, looking at the competing power relations involved, and asking who wins, and who loses in policy formulation and its implementation. Finally, areas of contention and debate will be identified for each country setting, highlighting areas for institutional and policy development (for example, around issues of regulation, certification, priority setting and so on).

SEED MULTIPLICATION

Governments in the region noted about 30 years or so ago that the disparity between food production and population increase in Africa south of the Sahara was ever widening with a resultant decrease in per capita food consumption. Due to population increase land availability was declining. The best approach to increasing food production was, therefore, to increase crop yields. But farmers were growing mostly their own traditional varieties, which are low yielding. Some high yielding varieties had been released but were not available to farmers. As a result quasi-government seed companies were established in Malawi (National Seed Company of Malawi, NSCM) and Tanzania (TANSEED). After a few years of operation, these companies abandoned their self-pollinating crop seed enterprises citing poor profitability. As a result, low adoption of recommended technologies for beans still remains a major problem in the region despite the development of new high yielding varieties. Such barriers to seed dissemination are a major constraint to the determination of impact from CRSP programs. The East Africa Bean/Cowpea CRSP Project therefore embarked on development of locations and means to multiply and disseminate basic and foundation seed of improved varieties.

Earlier studies at Bunda indicated that growing beans in Malawi during the hot dry season acts as a filtration process of seedborne diseases except viruses. As a result, the programme embarked on increasing breeders' seed under irrigation on campus. In each of the past five years breeders' seed has been increased to supply basic seed. At SUA basic seed is also produced similarly. During the past five years the Malawi Bunda CRSP multiplied basic seed of Kalima on about 5ha in each year to produce foundation seed. This seed was sold to NGOs, *e.g.*, ActionAid and Self-Help Development International, and to Ministry of Agriculture for demonstrations. In Tanzania the bean seed distribution system is significantly different from that in Malawi. Until recently, bean seed was produced exclusively on government farms with little seed being produced or reaching the farmer. As the shortcomings of the official Ministry of Agriculture multiplication and dissemination stations become apparent and as market liberalization has progressed, more decentralized seed production systems are developing.

Barriers to seed dissemination result because of a number of socio-economic constraints. Poor distribution of inputs and produce in the region results from poor infrastructure that exists. The road/bridge network is so vulnerable that vast areas are inaccessible particularly during the rainy season. Markets are not adequately established. The quasi-government marketing organizations have better networks of markets but are poorly organized. They offer the farmers the lowest prices and generally do not have ready cash to offer the farmer when the crop is ready for sale. Entrepreneurs are coming into the system due to market liberalization. They may have ready cash and transport but lack necessary knowledge to handle seeds differently from other inputs and produce. In many instances farmers are not organized into groups or cooperative societies, which are powerful bodies that would assist them do combined efforts to break through the barriers of the marketing systems. In Malawi the credits system broke down as the country changed from dictatorship to a plural society.

It has been realized that smallscale bean farmers will:

- Buy bean seed and not just rely on their own stocks,
- Afford to buy seed of new varieties, and
- Not function efficiently in varietal diffusion.

Non-governmental organizations are increasingly filling a niche in seed dissemination activities and thereby providing temporary solutions to the barriers expressed previously and to facilitate farmers' capabilities for seed production. Their usual goals are food security and poverty alleviation. Malawi has a greater profusion of NGOs because of its burden of Mozambican refugees in the 80's and 90's. These NGOs have stayed even after most refugees were repatriated. In general, these NGOs operate two types of seed schemes; *viz.*: commercially-oriented schemes where fewer farmers purchase more seed to produce seed for sale and the food security oriented schemes where more farmers are given less seed on credit to produce seed or grain for the market or food. The former crop is certified whereas the latter is hardly inspected but the seed is 'approved'. Approved seed is that seed which is produced either from certified or known stocks of uncertified seed. Seeds of this category do not necessarily carry a government seal but have to pass seed standards for purity and germination. This has facilitated the operations of many NGOs in seed multiplication and dissemination in Malawi.

In Malawi two organizations involved in seed distribution are described. The first is an NGO (ActionAid) that has a food security-oriented seed scheme. ActionAid developed a 'Malawi Smallholder Seed Development Project, MSSDP' after a flood disaster in 1992 when 1.1 million households benefited from 3,000 tonnes of relief seed issues.

The objectives of this Project are to:

- Develop low cost mechanisms, which would improve and sustain availability and accessibility of appropriate, improved seed to resource poor farmers in participating communities,
- Establish sustainable self-motivating community based groups, which will manage seed multiplication and distribution in the communities, and
- Train selected extension staff and participating community groups in seed production, seed quality control, on farm seed selection and storage, group dynamics and in community participatory methodologies.

In 1998 a total of 3.6 tonnes of certified bean seed was supplied and made available to 76 groups with a total membership of 1500, who in turn multiplied more than 12 tonnes of first generation approved bean seed. About 75 per cent of this seed were Kalima and Nasaka from CRSP.

An example of the commercially oriented seed scheme is that operated by the Maize Productivity Task Force, formed in Malawi in 1995, with the aim of increasing agricultural productivity. Its Action Group Two, one of the four groups, was set up to increase availability of seeds of self-pollinating and open pollinated crop varieties, which are neglected by the large multilateral seed companies. Funded by European Union, this group embarked on seed production and marketing as a business.

The major activities of this group are:

- Technology promotion through demonstrations, radio programs and field days;
- Training of extension staff and seed growers in seed production techniques;
- Organization of farmers through establishment of farmer associations; and
- Seed production.

In 1998/99 season there were 109 farmers growing seed beans on 28ha producing 24 tonnes of seed.

The CRSP project in Tanzania released SUA90 and Rojo. These varieties have been introduced into some selected farming communities in the country. The areas where they are currently grown and marketed include Morogoro and Kilosa districts of Morogoro region, Kongwa district of Dodoma region and some areas of Tanga and Arusha regions. Seed of Rojo was recently produced at SUA, Tanzania and was purposely disseminated through two NGOs, *viz.*: LVIA (Lay Volunteers International Agency) and CCT (Christian Council of Tanzania) and through the CRSP Project. The CRSP Project distributed the seed to Kongwa, Kisanga and Msolwa villages in addition to farmers in Maharaka and Msongozi, Muhenda and Ulaya-Mbuyuni villages of Morogoro and Kilosa districts. LVIA selected participating farmers based on their willingness to engage in seed production and on financial ability to start seed production. For SUA CRSP gender and education were major criteria. CCT selected progressive and innovative farmers in terms of knowledge on improved agricultural production. Notable amongst the results is the fact that smallholder farmers are willing to purchase foundation seed. However, seed demand and marketing issues are clearly not addressed. Farmers were unable to sell their seeds within their villages and had to sell as cheap grain. Adequate extension is lacking. The majority of farmers are still unaware of the new seed. LVIA provided its own extension staff and farmers under their scheme were more knowledgeable. Seed promotion activities also worked better for farmers under their scheme. A general recommendation was that successful seed producers are those more financially endowed and that these are the farmers to be targeted in future.

DISEASE MANAGEMENT

Black rot management begins with the identification of potential disease sources and utilising an Integrated Pest Management (IPM) strategy including host resistance, planting disease free seed, avoiding spreading the disease and proper sanitation. Sanitation is the main method that reduces, excludes or eliminates the initial sources of disease. General sanitation practices include crop rotation, disinfecting seed, rouging diseased plants, elimination of refuse piles and eradication of alternative hosts.

SEED TREATMENT

Seedborne inoculum significantly contributes to the spread of black rot bacteria. Growers should only plant tested certified seed < 1 infected seed in 30,000 or 0.003 per cent contamination. When the infection level of seed is not known or disease-free seed is not available, seed should be treated to eliminate the bacteria. Growers who purchase transplants should request proof the seedlings were grown from disease-free or treated seed. During transplanting, diseased seedlings should not be planted in the field. Seed treatments do not always eliminate 100 per cent of the bacteria on or in the seed, and may adversely affect seed germination and vigour. Soaking seeds in hot water at 50 °C for 25–30 min. is the most effective treatment for seedborne blackrot control. Weak seed, seed stored for several years and seed of certain crucifer crops; such as, cauliflower, kohlrabi, kale, rutabaga and summer turnip, may be damaged by hot water treatment; soak for 15 min. at 50 °C only.

The effect of the hot water seed treatments on every variety of each individual crucifer crop has not been investigated. Growers are encouraged to treat a small portion of seed and plant in pots to determine the effect of the seed treatment on germination and vigour, prior to treating the entire seed lot.

AVOID DISEASE SPREAD

Use new seed trays each year to avoid contaminating this year's crop with residual black rot bacteria from the previous year. If purchasing new trays each year is not economically feasible, used trays can be sterilised with steam, boiling water or chemical disinfectants to eliminate potential contamination. Destroy infected seed trays immediately to prevent disease spread to other seedling trays. Avoid soaking crates or bundles of transplant seedlings in tubs of water before transplanting. The black rot bacteria can spread from diseased to healthy seedlings by infecting leaf scars and wounds on roots when soaked in water.

Black rot bacteria can contaminate the surface of clothing, equipment, tools and water sources. Reducing seeding rates and densities to promote good air circulation, facilitating the quick drying of plants, timing irrigation when plants will dry quickly and restricting field activities until later in the day when fields are dry will help reduce disease spread. Working in diseased fields last will also avoid disease spread from infected to non-infected fields. Wash and disinfect equipment before moving from one field to another.

FIELD SELECTION

Field selection is very important due to the distance the pathogen can spread. Whenever possible, select fields as far away from fields grown to crucifer crops the previous year. Select fields that are well drained and will not receive run-off water from areas or fields where crucifers have been grown previously. Well drained, light soils are best for crucifer production because they can be worked early in the season and facilitate earlier planting of transplants. Planting early can help avoid disease because environmental conditions are usually not conducive for the development and spread of black rot bacteria.

INSECT CONTROL

The crucifer flea beetle (*Phyllotreta cruciferae*) can transmit black rot bacteria from infected plants to healthy ones; however, their importance in the spread of the disease is limited. Wounds caused by insects provide an entry point for the disease to infect plants during heavy dews or periods of rain. Insect control will help reduce the spread and severity of disease.

CULL PILE MANAGEMENT

Infected refuse or cull piles left in the field, provides an excellent source of the black rot bacteria. Fresh cull piles left near fields can result in severe disease epidemics during the growing season. Prepare cole crops for market away from fields, and immediately chop and bury the diseased tissue cut from plants.

RESISTANT VARIETIES

The development of crop varieties with disease resistance or tolerance to black rot has been the focus of many cole crop breeding programs worldwide. Resistance to black rot was first identified in the Japanese cabbage cultivar, Early Fuji. Today, many crucifer hybrids with black rot tolerance are available for both fresh and processing commercial production.

CHEMICAL CONTROL

Soil fumigation can significantly reduce black rot bacteria. Soil fumigation is expensive and alternative methods for managing plant pathogenic bacteria are needed. For more information on chemical control options refer to OMAFRA Publication 363, *Vegetable Production Recommendations*.

CROP NUTRITION

The effect of plant nutrient management on the susceptibility of host crops to black rot infection is not fully understood. A balanced nutrient programme may reduce the susceptibility of plants to disease infection.

Excess nitrogen promotes lush vegetative growth and may increase plant susceptibility. Micronutrients may also be involved with the disease defence mechanisms of crucifer crops.

CROP ROTATION

Planting disease-free, treated seed or seedling transplants does not necessarily ensure a disease free crop in the field. Crop rotation is also an important management tool. Black rot bacteria can survive in infected crop tissue in soil until the crop tissue breaks down and rots. The time required for crucifer crop debris to rot varies between regions depending on the temperature, amount of soil moisture and soil type. For example, in the states of Georgia and Washington, which experience long, warm summers, it has been estimated that free-living bacteria can survive in infested soil for about 60 days, and up to 615 days in infested host debris. The bacteria can survive longer in soil during cool, wet seasons than during hot, dry seasons. In Ontario, a 3-year rotation is recommended.

WEED CONTROL

Black rot bacteria can infect and survive on many crucifer weeds including bird rape (*Brassica campestris*), Indian mustard (*B. juncea*), black mustard (*B. nigra*), shepherd's purse (*Capsella bursa-pastoris*), globe-podded hoary cress (*Cardaria pubescens*), pepper grass (*Lepidium densiflore*) and wild radish (*Raphanus raphanistrum*). Disease symptoms on weeds vary from small yellow V-shaped lesions on leaf margins to no visible symptoms. The pathogen can spread up to 30 m from infected plants (including weed hosts) to healthy plants. The pathogen not only infects and spreads from weeds to cruciferous crops, it can also survive on weed seeds and can grow and multiply on weed leaves without infecting or causing disease. Good weed control within fields will aid disease management; however, careful attention to weed control in ditches and along fencerows is also important.

TECHNICALITY IN PRODUCTION

REQUIREMENT OF SEED PRODUCTION

The basic requirement of seed production is availability of improved varieties/hybrids and their demand among the farmers. Preferably the variety should be realsed and notified for certified seed production, however, any kind/variety can be multiplied for Truthfully Labeled Seed (TFL).

ISOLATION REQUIREMENTS

The cucurbits are cross pollinated in nature and honeybees are major pollinator, thus for pure seed production an isolation distance all around seed field is necessaryto separate it from fields of other varieties, fields of the same variety not confirming to varietal purity requirement. The isolation distance of 400 m for C.S. and 800 m for F.S. and at least1000 m isolation is required for breeder seed production.

It is important to mention that muskmelon, longmelon and snapemelon (phoot) can cross with each other. Similarly, species of genus Cucurbita have the risk of crossability among them. The cucumber can cross easily with its wild relative Cucumis hardwickee found in wild form in sub-mountanious regions of Himalayas.

CHOICE OF SEASON AND AREAS OF SEED PRODUCTION

Seed crop should be raised in such a seasons which remain dry at the time of seed maturity and seed extraction. Rainy season is preferred over summer season for raising seed crop.

Locations are also important in seed production with reference to seed yield and quality of seed. To harness the advantage of climate, private sector seed companies are organizing their seed production in these areas.

MAJOR CUCURBITS SEED PRODUCTION REGIONS IN INDIA

Crop	**Areas**	**Region**
Muskmelon, Longmelon	Punjab & Haryana	Northern Region
Bittergourd	Eastern U.P., Faizabad and Jaunpur	Northern Region
Cucumber and Muskmelon	Azamgarh, Ballia and Gonda in U.P. Jalana (Maharashtra)	Northern Region Central Region
Pumpkin and Ridge gourd	West Bengal (South West)	Eastern Region
Sponge gourd, Watermelon, Cucumber	Telingana, Karool and Vijayawada (North A.P.) Ranibenur (Karnataka)	Southern Region Southern Region
Watermelon, Bottle gourd and Bitter gourd	Costal districts of A.P.	Southern Region

Roguing

Seed crop is to be monitored at various stages of crop growth for removal of off-type and obviously should be carried out before flowering to avoid natural cross-pollination.

However, fruit set and complete fruit development stages are also important. The crop-wise main features described here under for effective roguing.

- *Muskmelon:* Fruit shape, colour, rind colour, skin (netted/plain), flesh colour (orange/red), TSS and cavity size.
- *Watermelon*: Fruit shape, colour, rind colour, flescolou(red/yellow/ white).
- *Longmelon*: Fruit shape, colour, bitterness.
- *Cucumber*: Fruit shape, colour, presence of spines, spines colour, colour of ripen fruit (green, yellow, white or orange)
- *Pumpkin (Squash, summer)*: Fruit shape, colour, flesh colour.
- *Gourds (Bottle gourd, Bitter gourd and Luffa etc.)*: Fruit shape, colour, stripe, neck etc.

MATURITY OF FRUIT

Cucurbits takes fairly long time to attain harvestable maturity. The maximum period is required in crops like pumpkin (Cucurbita moschata), ashgourd and watermelon, however, muskmelon, round melon and bitter gourd take relatively lesstime. The maturity also influenced by the environmental factors and crop management (trailing etc.). The maturity period is shorter in summer season than rainy season.

HARVESTABLE MATURITY IN CUCURBITS IN INDIA

Besides days to maturity, some other parameters like change in colour also used as crietaria of maturity index.

- *Cucumber and summer squash*: Fruit turn pale yellow to golden yellow and attached with plant.
- *Pumpkin*: Fruit reddena and seeds inside the shell breake readily from pulp.
- *Muskmelon*: Full slip stage.
- *Watermelon*: Fruits are ready for harvest when they reach edible maturity, fruit colour change from green/white to pale yellow of under side of the fruit.
- *Bitter gourd and snake gourd*: Fruit turn to bright yellow.
- *Bottle gourd*: At maturity fruit colour fade to starw freen or pale yellow.
- *Luffa*: Complete drying/fruit turn to gray colour.

Crop	Variety	Period in Days (Seed to Seed)
Bitter gourd	Priya, Pusa Vishesh	65 days
Snake gourd	TA-19	65 days
Bottle gourd	Arka Bahar,	75 days

	Pusa Naveen	80 days
Pumpkin	Ambali	90 days
	Pusa	(Rainy season)
	Vishwas/Vikas	100 days
Ash gourd	Local	90 days
Cucumber	Mudikode Local	70 days
Watermelon	Sugar Baby	85 days
Muskmelon	Pusa Madhuras	80 days

SEED EXTRACTION

There are two method of seed extraction employed in cucurbits:

1. *Dry method*: The dried fruits are cut from one side and the seeds comes out from the fruit *e.g.,* sponge gourd, ridge gourd, snake gourd.
2. *Wet method*: This method is employed for seeds extraction of cucumber, muskmelon, watermelon, ash gourd, bitter gourd, round melon and long melon. The fruit of cucumber and bitter gourd, summer squash and long melon are cut longitudianlly and seed is scooped out while fruit of muskmelon and pumpkin are cut into two piece and seed is scooped out from cavity. However, in case of watermelon and ash gourd whole central portion are manually scooped out and macerated to separate the seed from pulp. In wet method, the seed extraction done by three ways:
   - *Mechanical Extraction:* In this method the fruits are cut into pieces and macerated by machine. The seeds are separated out from pulp by floating with water. This method is quick, less expensive and seeds retain good lusture, but require good amount of water. This method is applicable in bottlgourd, watermelon, roundmelon and ash gourd.
   - *Natural Fermentation*: The scooped material kept in wooden/ plastic or steel vesel for 48 hours at room temperature and stired 2-3 times and then seed is washed throughly with water 2-3 times. The main problem with this method are discolouration and poor lusture of seed.
   - *Chemical Extraction*: 25-30 ml. of HCL or 8-10 ml. of commercial H_2SO_4 added per 5 kg of pulp and some quantity of water is mixed, stiring of pulp is done to enhance to separation and left for 30 minutes. The impurities will float and seed will sink. The seed should be washed throughly with clean water. This is quick method but accuracy of acid and time is important.

CONQUEST OF PRODUCTION

Even today we see men and women denying themselves necessaries to acquire mere trifles, to obtain some particular gratification, or some intellectual or material enjoyment. A Christian or an ascetic may disapprove of these desires for luxury; but it is precisely these trifles that break the monotony of existence and make it agreeable. Would life, with all its inevitable sorrows, be worth living, if besides daily work man could never obtain a single pleasure according to his individual tastes?

If we wish for a Social Revolution, it is no doubt in the first place to give bread to all; to transform this execrable society, in which we can every day see robust workmen dangling their arms for want of an employer who will exploit them; women and children wandering shelterless at night; whole families reduced to dry bread; men, women, and children dying for want of care and even for want of food. It is to put an end to these iniquities that we rebel.

But we expect more from the Revolution. We see that the worker compelled to struggle painfully for bare existence, is reduced to ignorance of these higher delights, the highest within man's reach, of science, and especially of scientific discovery; of art, and especially of artistic creation. It is in order to obtain these joys for all, which are now reserved to a few; in order to give leisure and the possibility of developing intellectual capacities, that the social revolution must guarantee daily bread to all. After bread has been secured, leisure is the supreme aim.

No doubt, nowadays, when hundreds and thousands of human beings are in need of bread, coal, clothing, and shelter, luxury is a crime; to satisfy it the worker's child must go without bread! But in a society in which all can eat sufficiently the needs which we consider luxuries to- day will be the more keenly felt. And as all men do not and cannot resemble one another (the variety of tastes and needs is the chief guarantee of human progress) there will always be, and it is desirable that there should always be, men and women whose desire will go beyond those of ordinary individuals in some particular direction.

Everybody does not need a telescope, because, even if learning were general, there are people who prefer examining things through a microscope to studying the starry heavens. Some like statues, some pictures. A particular individual has no other ambition than to possess an excellent piano, while another is pleased with an accordion. The tastes vary, but the artistic needs exist in all. In our present, poor capitalistic society, the man who has artistic needs cannot satisfy them unless he is heir to a large fortune, or by dint of hard work appropriates to himself an intellectual capital which will enable him to take up a liberal profession. Still he cherishes the *hope* of some day satisfying his tastes more of less, and for this reason he reproaches the idealist Communist societies with having the material life of each individual as their sole aim.— "In your communal stores you may perhaps have bread for all," he says to us, "but you will not have beautiful pictures, optical instruments, luxurious furniture, artistic jewelry—in short, the many things that minister to the infinite variety of human tastes. And in this way you suppress the possibility of obtaining anything besides the bread and meat which the commune can offer to all, and the grey linen in which all your lady citizens will be dressed".

These are the objections which all communist systems have to consider, and which the founders of new societies, established in American deserts, never understood. They believed that if the community could procure sufficient cloth to dress all its members, a music hall in which the "brothers" could strum a piece of music, or act a play from time to time, it was enough.

They forgot that the feeling for art existed in the agriculturist as well as in the burgher, and, notwithstanding that the expression of artistic feeling varies according to the difference in culture, in the main it remains the same. In vain did the community guarantee the common necessaries of life, in vain did it suppress all education that would tend to develop individuality, in vain did it eliminate all reading save the Bible. Individual tastes broke forth, and caused general discontent; quarrels arose when somebody proposed to buy a piano or scientific instruments; and the elements of progress flagged. The society could only exist on condition that it crushed all individual feeling, all artistic tendency, and all development.

Will the anarchist Commune be impelled by the same direction? Evidently not, if it understands that while it produces all that is necessary to material life, it must also strive to satisfy all manifestations of the human mind.

SEED YIELD

Seed yield depends upon the crop, variety, location, season and management of the seed crops. Seed attributes of selected cucurbits varieties of IARI.

Crop	Variety (gm)	Seed Yield/ (gm) wt.(gm)		Fruit Seed/Fruit	No. of Seed	1000
		Summer Season	Rainy Season			
Bottle gourd	Pusa Naveen	63.00	73.33	634.93 and 654.69 in summer and rainy season	120-130 gm	
Sponge	P. Supriya	25.78	–	248.44	–	
gourd	P. Chikni	–	9.93	–	–	

Pumpkin	P. Vikas	52.83	–	270.00	200 gm
Bitter gourd	P. Vishesh	4.052	–	21.77	160-170 gm

ROLE OF SOCIETY

Is it a dream to concieve a society in which—all having become producers, all having received an education that enables them to cultivate science or art, and all having leisure to do so—men would combine to publish the works of their choice, by contributing each his share of manual work? We have already hundreds of learned, literary, and other societies; and these societies are nothing but voluntary groups of men, interested in certain branches of learning, and associated for the purpose of publishing their works. The authors who write for the periodicals of these societies are not paid, and the periodicals are not for sale; they are sent gratis to all quarters of the globe, to other societies, cultivating the same branches of learning.

This member of the society may insert in its review a one-page note summarizing his observations; another may publish therein an extensive work, the results of long years of study; while others will confine themselves to consulting the review as a startingpoint for further research. It does not matter: all these authors and readers are associated for the production of works in which all of them take an interest. It is true that a learned society, like the individual author, goes to a printing office where workmen are engaged to do the printing. Nowadays, those who belong to the learned societies despise manual labour; which indeed is carried on under very bad conditions; but a community which would give a generous philosophic and *scientific* education to all its members, would know how to organize manual labour in such a way that it would be the pride of humanity. Its learned societies would become associations of explorers, lovers of science, and workers—all knowing a manual trade and all interested in science.

If, for example, the society is studying geology, all will contribute to the exploration of the earth's strata; each member will take his share in research, and ten thousand observers where we have now only a hundred, will do more in a year than we can do in twenty years. And when their works are to be published, ten thousand men and women, skilled in different trades, will be ready to draw maps, engrave designs, compose, and print the books. With gladness will they give their leisure—in summer to exploration in winter to indoor work And when their works appear, they will find not only a hundred, but ten thousand readers interested in their common work.

This is the direction in which progress is already moving. Even to-day, when England felt the need of a complete dictionary of the English language, the birth of a Littré, who would devote his life to this work, was not waited for. Volunteers were appealed to, and a thousand men offered their services, spontaneously and gratuitously, to ransack the libraries, to take notes, and to accomplish in a few years a work which one man could not complete in his lifetime. In all branches of human intelligence the same spirit is breaking forth, and we should have a very limited knowledge of humanity could we not guess that the future is announcing itself in such tentative co-operation, which is gradually taking the place of individual work. For this dictionary to be a really collective work, it would have required that many volunteer authors, printers and printers' readers should have worked in common; but something in this direction is done already in the Socialist Press, which offers us examples of manual and intellectual work combined. It happens in our newspapers that a Socialist author composes in lead his own article. True, such attempts are rare, but they indicate in which direction evolution is going.

They show the road of liberty. In future, when a man will have something useful to say-a word that goes beyond the thoughts of his century, he will not have to look for an editor who might advance the necessary capital. He will look for collaborators among those who know the printing trade, and who approve the idea of his new work. Literature and journalism will cease to be a means of money-making and living at the cost of others. But is there any one who knows literature and journalism from within, and who does not ardently desire that literature should at last be able to free itself from those who formerly protected it, and who now exploit it, and from the multitude which with rare exceptions pays it in proportion to its mediocrity, or to the ease with which it adapts itself to the bad taste of the greater number? Letters and science will only take their proper place in the work of human development when, freed from all mercenary bondage, they will be exclusively cultivated by those that love them, and for those that love them.

QUALITY SEED PRODUCTION

PARENTAL LINES FOR SOME OF THE HYBRIDS

Hybrids	Female Line	Male Line
1. CSH – 5	2077 – A	CS – 3541
2. CSH – 9	296 - A	CS – 3541
3. Kovilpatti tall	2219 – A	IS – 3541
4. COH – 3	2077 – A	699 Tall
5. CSH – 15 R	104 – A	R – 585
6. 296 – A	296 – A	TS – 30

In sorghum hybrid seed production, cytoplasmic genic male sterility system is employed. Here, the female plants are male sterile and called as 'A' line plants. The male plants are called at 'R' line or Restore line since it restores the male fertility in the F1 hybrid. In order to maintain the 'A' line plants, they are crossed with 'B' line or maintainer line plants. 'B' line plants are fertile counterparts of 'A' line plants. They are isogenic to 'A' line plants in all features except male sterility.

SELECTING THE RIGHT SEASON

For hybrid seed production in sorghum, sowing of parental liens can be tken up from October to December.

If flowering period coincides with dry and low cool conditions, not only seed set but also seed quality will be more. In order to achieve synchronisation of flowering between the parental lines, staggered sowing of the parental seeds should be followed.

Seed Rate

Female parental line – 7.5 kg ha-1
Male parental lines – 5 kg ha-1

Pre-sowing Seed Treatment

Hybrid sorghum seeds can also be produced under rainfed condition. For that, seeds should be infused with drought resistance by giving seed hardening treatment.

SEED HARDENING

Soak the seeds in 2 per cent potassium dihydrogen phosphate for 10 hours and then dry the seeds to the original moisture content. Alternately, botanicals like Prosopis juliflora can also be used for seed hardening purpose.

Botanicals are also eco friendly. In this method, seeds are soaked in 1 per cent Prosopis leaf extract at seed to solution ratio of 1:0:6 for 16 hours. Then the seeds are dreid to original moisture content. Dried seed scan be pelleted with dried pungam leaf powder @ 500 gm per kg of seeds.

To facilitate the leaf powder to stick on the seeds, 10 per cent maida solution can be used as adhesive. The plants arising from hardened seeds will withstand drought condition, if any, during the crop growth period.

Isolation Distance

Hybrid sorghum seed production filed has to be isolated from other sorghum crops atleast 200m.

Spacing

Before sowing, ridges and furrows with a spacing of 45 cm between the furrows are formed. Sow the seeds at 1/3 height of the ridge form bottom with a plant to plant spacing of 15 cm.

Four numbers of border rows should be sown with male line seeds all around the seed production field. Then, female and male line seeds are to be sown with a ratio of 5:2 which means that five rows of female line should be alternated with two rows of male. This is called planting ratio.

Border Rows

During hybrid seed production, four rows of male parental lines are to be grown all around the field. They are called as border rows. The border rows will provide pollen grains to the female plants.

Fertilizer Application

Fertilizers at the rate of 100:50:50 kg Nitrogen, Phosphorus and Potash are to be applied per ha. Of this, 50:50:50 kg of NPK is applied as basal.

First top dressing with 29 kg of N is to be applied after first weeding. Second dose of 25 kg N can be given 45 days after sowing.

Micronutrient Deficiency

If micronutrient deficiency occurs, not only the seed yield is reduced but also the seed quality will be affected:

- Zinc sulphate at 25 kg per ha may be applied to the zinc deficient soils.
- For iron deficiency, 50 kg of ferrous sulphate with 12.5 tonnes of farm yard manure is to be applied to the soil.
- During plant growth stags, intervenal chlorosis may occur due to iron deficiency. For that, 0.5 per cent ferrous sulphate should be applied as foliar spray.

Roguing

Roguing operation involves the removal of offtype plants from the seed field. To maintain genetic purity, offtypes should be removed before flowering stage. In a hybrid seed crop, it is important to carry out rouging both in male and female rows. Besides this, some of the pollen shedding plants called as pollen shedders should also be removed from the female rows. Farm women will attend the rouging operations in an effective way.

What is Offtypes?

Plant or seed deviating significantly from the characteristics of a variety as described by the breeder in any observed respect.

Attention

Pollen shedders have to be removed and carried away from the early stages. For weed control, 200 gm Atrazine dissolved in 1000 litre of water for one ha can be sprayed. After 30-35 days, hand weeding is necessary.

SYNCHRONISATION OF FLOWERING

In sorghum hybrid seed production, the female line used is male sterile. For fertilizing the flowers of the female line, pollen has to come from the male plants. If only the flowering in the female line coincides with the flowering in the male line, proper pollination and seed set in the female are possible. Simultaneous flowering in both female and male plants is called as synchronisation of flowering.

Panicle Initiation in Sorghum

Methods to Achieve Synchronisation

Best method to achieve synchronisation is to stagger the sowing of parental lines based on previous knowledge about the duration of flowering. Late parent should be sown early and early parent late.

In spite of this precaution, if a gap in flowering period is predicted based on panicle initiation, then the following methods can be followed to hasten the late parent or delay the early parent by 2 or 3 days:

- The advancing parent has to be sprayed with 500 mg malic hydrazide in one litre of water, 45 days after sowing. Maleic hydrazide will not dissolve in water, you have to dissolve it in a little volume of sodium hydroxide and then make up the volume with required quantity of water.
- 1 per cent urea solution has to be sprayed on the late parent.
- One irrigation may be skipped for the advancing parent.
- Flowering can be delayed, by spraying CCC (Chloro choline chloride) at a concentration of 300 ppm.

Attention

High yield of hybrid seed can be obtained only when proper synchronisation of flowering occurs between the female and male plants.

Stage of Harvest

Harvest the crop 40-45 days after 50 per cent flowering. At that time, seed moisture would be around 25-30 per cent. A black layer will be visible at the base of the seed is the visible symptom of seed maturity.

Harvesting

First, the male rows including the border rows are to be harvested and the earheads removed from the field. Then, harvest the female rows. Seeds harvested from the female rows are the hybrid seeds.

Seed Yield

1500 kg ha-1

Threshing

Seeds can be threshed either by manual method or by using mechanical threshers. During threshing, if seed moisture is kept at 15-18 per cent, mechanical injury to the seed will be less.

Seed Drying

Seed can be dried either under sun or by using mechanical driers. If you dry the seeds under sun, avoid the noon hot sun. During the noon time, temperature will be very high and exposing the seeds to such high temperature will affect viability. If mechanical drier is used, ensure that the drying temperature does not exceed 40°C.

SEED PROCESSING

Process the seeds with 9/64" round perforated sieve for size grading.

Seed Storage Methods

Before seed storage, seed has to be dried to 12 per cent moisture content. Seed treatment with Thiram or Captan 75 per cent wp at the rate of 70 gm in 500 ml of water per 100 kg seeds is necessary. After this treatment, seeds can be stored for more than one year. Seeds dried to 8 per cent moisture content can be stored upto one and half years in moisture vapour poof containers like polythene bags.

Seed Certification

Seed certification guarantees the quality of seed as it ensures that the certified seed has the genetic, physical, physiological and seed health qualities.

Genetic purity means that the seed gives rise to a plant which conforms to the varietal characteristics of the variety. The physical purity means that the seed is free from stones, broken seeds, straw bits and leaf bits etc. Physiological quality is measured by germination and seed health envisages freedom from pest and diseases.

Seed certification is being done in many stages. It starts from verifying whether seeds were obtained from authenticated source, verification of isolation distance and inspection during plant growth, flowering, harvesting, processing and bagging. Also seed samples are drawn form the seed lot and sent to seed testing lab to test whether the seeds are possessing required physical purity and germination.

Then certification tag is issued. Colour of the tag is blue for certified seeds. Only those seeds harvested from fields having prescribed field standards and possessing required seed standards are certified by the Certification Agency. Seeds thus certified are offered for sales.

Field Standards (Certified Seed)

	Factor	**Maximum Permitted per cent**
1.	Off types (per cent) (maximum)	0.10
2.	Pollen shedders (per cent) (maximum)	0.10
3.	Ergot affected plants (maximum)	0.10

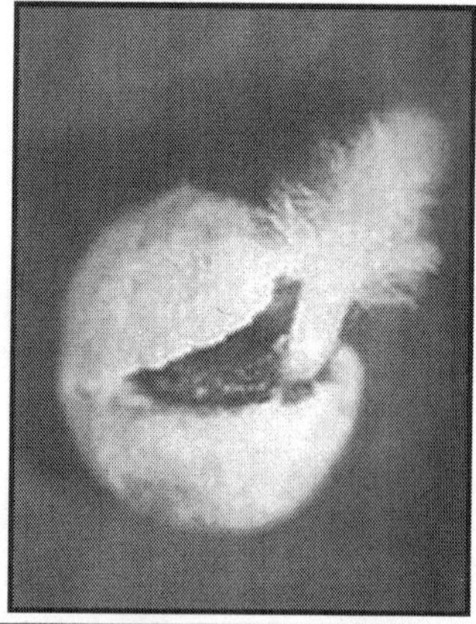

Seed Standards (Certified Seed)

	Factor	**Standard**
1.	Physical purity (per cent) minimum	98.0
2.	Inert matter (per cent) maximum	2.0
3.	Others crop seeds (maximum)	10 Nos/Kg
4.	Weed seeds (maximum)	10 Nos/Kg
5.	ODV (maximum)	20 Nos/Kg
6.	Germination (per cent) (minimum)	75

7.	Seed Moisture (per cent) (maximum)	
	Moisture pervious container	12.0
	Moisture impervious container	8.0

CONTRACT FARMING IN SEED PRODUCTION

Seed is important in sunflower production. However, quality seeds are not available to meet the growing areas of sunflower in Karnataka. To meet the sunflower seeds, KSSC,KOF, UAS etc. agencies are involved in seed production. The seed production targets are being fulfilled by several ways including contract farming. However, there are several issues raised by different stakeholders about the profitability and feasibility of seed production under contract farming system. Hence, the present study was carried out to address these issues.

Generally, there are three types of contract in agriculture viz."

1. Procurement contracts, under which only sale and purchase conditions are specified
2. Partial contracts, wherein only the contracting firms supply some of the inputs and produce is bought at preagreed prices and
3. Total contracts, under which the contracting firm supplies and manages all the inputs on the farm and farmer is just a supplier of land and labour.

The relevance and importance of each type varies from product to product and these types are not mutually exclusive. Whereas, the first type is generally referred to as marketing contracts, the other two are of production contracts. But, there is a systematic link between product market and factor markets under the contract arrangements as contracts requires a definite quality of produce. Different types of production contracts allocate production and market risks between the producer and the processor in different ways.

The scenario of contract farming can be traced to sugar mills in Karnataka were practicing the contract farming since many decades, where farmers were growing sugarcane at the specified pre-agreed price. In the same manner, corporate sector has introduced it to horticultural and vegetable crops. Similarly, the contract farming of oilseeds for seed production was introduced in Northern Karnataka during 1985-86. It was first introduced in the taluks of Haveri, Ranebennur, Hanagal and Hirekerur of Haveri district by the different firms like KOF, KSSC and private seed companies *viz.*, Mahyco and Sungro. The farmers of this district had engaged since long time in the production of oilseed for seed purpose.

IMPORTANCE OF THE PRESENT STUDY

The KOF in seed production resort to contract farming mainly to have assured supply of genuine seed material in required quantity at the right time, which has been produced under their supervision. On the other hand, the farmers are interested to enter into contract mainly to minimize the price risk, and also to reap higher profits out of this seed production activity over commercial production of crops.

The present investigation is an integrated effort to study all socio-economic aspects and profitability of production of hybrid sunflower seeds in addition to identifying the constraints in its production with an overall view of exploring the possibility of bringing about required improvement. The results obtained from this study would be used to overcome the present limitations in production of hybrid sunflower seeds.

The information on cost and returns structure will guide the producer in readjustment and proper management of resources and to bring down the cost of production at the farm level without affecting the output. The results would help the planners and policy makers in formulating suitable policies for grant of loans and fixation of prices and also throw further light on the avenue for future research in the area of hybrid sunflower seed production.

PRODUCTIVITY OF RESOURCES IN SEED PRODUCTION

Kshirsagar studied the resource use efficiency of different inputs in hybrid cotton seed production using Cobb-Douglas production function. Out of four selected variables in Varalaxmi, the human labour input was significant at one per cent level, while fertilizers and plant protection measures were significant at five per cent level of probability. In the case of H-4 cotton seed, except fertilizer, all selected variables *viz.,* human labour, plant protection measures and irrigation were significant at five per cent level of probability. Patel studied the resource use efficiency of various inputs used in seed production of hybrid cotton in Gujarat. The ratios of MVPs of different factors to respective cost of factors were calculated. It was found that 9.14, 1.23, 5.48, -1.08 and 6.83 ratios of MVP to MFC of bullock labour, human labour, plant protection chemicals, irrigation and other expenditures, respectively.

Pathak studied economics of production and marketing of certified seeds of high yielding varieties of jowar and wheat. He concluded that the factor product relationship, between the output and human labour, bullock labour and manures and fertilizers was positive in jowar seed production. In wheat seed production, it was seen that there was a positive relationship between the output and human labour, manures and fertilizers. Muralidharan studied the resource use efficiency in rice production in Kerala employing the Cobb-Douglas type of production function. Adjusted R2 (0.84) indicated that per cent of the variation in yield of paddy could be explained by the estimated production function. The coefficient of land and human labour were positive and significant at one per cent probability level.

Radha *et al.*evaluated the resource-use efficiency in rice-rice and rice-pulse farming systems in Krishna district of Andhra Pradesh. The results indicated that manures and fertilizers and irrigation were quite productively used in both the farming systems. The sum of production elasticities indicated the operation of constant returns to scale in both farming systems.

Karisomanagoudar employed Cobb-Douglas type of production function in Gadag taluk of Dharwad district to study resource use efficiency in rainfed onion roduction. It was observed that land and labour inputs significantly increased the gross revenue. The seed variable exercised a significant negative influence on earnings from onion. The variables included in the production function explained 96 per cent of the variation in output.

Vasudha assessed the relationship between agricultural output and selected explanatory variables in Karnataka by using Cobb-Douglas production function. It was found that the gross cropped area, fertilizer consumption, bullock labour and human labour were important variables in explaining variation in agriculture output. The study also indicated that the agricultural production was increased at the rate of 3.11 per cent per annum during the period between 1956 and 1983 and fertilizer and irrigation were the predominant contributors for growth.

9

Important of Seed Dispersal

Saguaro seeds reach the ground in many ways. Some fall to the soil surface while they are still in the flesh of ripe, open fruits and in dried receptacles. Relatively heavy seedfalls occur during summer rainstorms. Rains soften and dislodge seeds from ripe fruits and dried receptacles remaining on the plant as well as from those that become lodged in the branches of associated trees and shrubs, and those already on the ground.

Summer rains aid importantly in transporting saguaro seeds to protected downslope sites affording both concealment from predators and favourable conditions for germination. After rains, seeds are found in the first few millimeters of the soil, in sand and gravel, under rock and in litter (usually mud-spattered), and nearly all are invisible to the naked eye. Natural germination occurs in such protected physical situations. Animals play a primary role in dispersal and the transport of seeds to sites suitable for germination and seedling establishment.

Most fruits drop to the ground, resulting in a high concentration of seeds at the base of the parent plant and providing an attractive supply of food for mammals, birds, and insects. The congregation of herbivorous predators1 not only results in low probabilities for seed survival but also severely limits seed germination and seedling establishment. Moreover, intensive digging, especially by rodents, effectively reduces protective plant cover at the base of the parent saguaro and destroys young saguaros or exposes them to other hazards. Generally, the probability of seed survival, germination, and seedling survival in creases with distance from adult plants. 1For lack of more appropriate terms we use the words "predator" and "predation" as broadly defined to include consumption and destruction of seeds and/or plants by animals.

Relatively low density populations of harvester ants (Pogonomyrmex sp.) and Harris' antelope ground squirrels (Citellus harrisi harrisi) are present, and round-tailed ground squirrels (Spermophilus tereticaudus) do not occur at this location. In the absence of heavy predation, a relatively large proportion of the annual seed cropremains on the ground until the occurrence of germinating rains.

Fig. Typical Accumulation of Ripe Fruits Beneath Mature Saguaros in Flat Habitats at the West Monument. On 9 July, with Six Ripe Fruits Remaining on the Plant, 179.5 g (6.3 oz)—approximately 131 Thousand Seeds—were Recovered from Detached Seed Masses and 52 Fruit Receptacles Surrounding the Base of this 3-branched, 9-m (29.5 ft) Tall Saguaro.

Fig. Typical Absence of Fruit at the Base of Mature Saguaros in Flat Habitats at the East Monument. Detached Seed Masses are Absent and not Seeds are Present in 45 Receptacles Remaining on the Ground Beneath this Vigourous 5-branched, 11-m (36.1-ft) Tall Plant. One Ripe Fruit Still Remains on the Plant.

Relatively dense populations of harvester ants (Pogonomyrmex sp.) and round-tailed ground squirrels (Spermophilus tereticaudus) consume most of the seeds within a few hours after the fruits drop to the ground. In this habitat, the area surrounding the base of an adult saguaro provides little or no protection—either for seeds or seedlings—from the intensive foraging of mammals, birds, and insects. Photographed 19 July 1971. In a matter of a few days, animal activity beneath the parent saguaro changes the situation from one resembling a fruit basket to the seed shadow associated with some tropical plant species as described by Janzen and Krebs. This explains why young saguaros are seldom found beneath the seed-producing parent plant. Those few seedlings that begin their establishment there do not survive the continuing heavy animal traffic at such stations. In addition to providing an important mechanism for seed dispersal, birds frequently deposit seeds in protected locations provided by trees and shrubs away from areas of intensive animal foraging at the base of fruiting plants. Such protected locations not only increase the probability of seed survival, they generally offer more suitable conditions for seed germination and seedling establishment.

AERIAL APPLICATION OF SEED

Aerial application of seed may be accomplished by using either of two basic types of contracts: Full Service or Call When Needed (CWN) or Job Contract.

Full Service or CWN Contracts - Under full service or CWN, the contractor is required to make specific aircraft available for exclusive use by the Government, normally for the transport of cargo or personnel. Payment is made by unit of time, generally hourly with a daily guarantee. The aircraft and pilot must be carded and the aircraft must be inspected by a FS/OAS mechanic. The benefit of this type of contract is that the Government exercises substantial control over the project. Specifications list all minimal requirements for aircraft and award based on a cost basis. This type of contract might be used where we are unsure of where treatment is needed on the ground. Because of the greater amount of control exercised by the Government, we also assume a larger portion of the liability and risk. The contractor doesn't have the same incentives to maintain an efficient operation which raises our cost risk. Risk of project failure is ours, along with liability assumed due to our directing the work.

JOB CONTRACT

Specifications in a job contract are written to state the work to be accomplished and the level of acceptability. As a minimum, we need to give potential contractors enough information to prepare a bid and perform the work under the contract. The following are factors which should be listed in the solicitation:

Locations and accessibility - Locations of project areas should be stated in terms of accessibility and written directions. They should be shown on maps. Information on elevation and terrain should also be provided as well as potential helispots and airstrips.

- Certification and licensing requirements.
- Government furnished property.
- Seed (if we provide), how it is packaged, size of container.
- Location of government furnished seed, how ft can be removed from storage, point contractor assumes control of seed.
- Topographic maps, orthophotos, airphotos to be furnished to contractor.
- Contractor furnished property and services.

- Seed (if they provide).
- Loading, transporting, and off-loading of seed.
- Selection of heliports/airstrips.
- Suitable aircraft with experienced pilots.
- Qualified ground supervisors and equipment.
- Dust abatement/warning signs/garbage cleanup.
- It should be stated that equipment moving and ferry time is incidental to cost of seeding.
- No government employees to ride in contractor furnished aircraft.

Avionics requirements should be listed. If the contractor must furnish any radios compatible with government radios, it must also be stated.

Seed mixes and application rates - Data tables should be included in the attachments that show seed mixes, application rates, and total pound by application block.

- Seed application system.
- Spreading system.
- Bucket type and size.
- Adjustable calibration.
- Calibration test requirements.
- Seed application.

List any criteria that will be used to determine when application operations will begin or cease:

- Wind velocity
- Fog/rain
- Surface run-off
- Air turbulence
- Darkness
- Equipment failure/avionics

List any maximum load limits:

- Altitude/airspeed requirements to meet calibration.
- Periodic checking of calibration.
- Application to cease during turns.
- No dumping of seed.

Measurement and payment - It should be stated how payment will be made. Normally payment will be based on pounds of seed satisfactorily applied within unit boundaries. Pound spread should be as determined in the inspection provisions. No payment should be made for seed not spread as required in the specifications.

Inspection and Maintenance

Avionics equipment should be inspected when the contractor arrives to ensure it is functional and satisfactory for intended

The rate of spread may be monitored in the following ways:

- By using seed collection devices.
- By calculating weight to acres.
- Observation of flight pattern and spacing as well as on the ground results.

Contract Time - A realistic amount of contract time should be allowed in the contract. In determining the contract time, consider the number of hours available during the normal work day when suitable weather will permit application to occur. This may only be 3-5 hours per day.

Even though a job contract may be used on an aerial seeding project, an approved aviation safety plan must be in place before the contract can be awarded. The contractor must comply with all Office of Safety and Health Administration (OSHA) safety regulations.

OPERATIONS PLAN

A detailed operations plan is a necessity for all aerial seeding projects. On such projects, a large amount of work needs to be accomplished in short time periods people are working in situations that may be considered dangerous, and people are asked to perform tasks that may be unfamiliar to them. Some people assigned to the incident may be from other administrative units and may not be familiar with the local geographic area or local administrative procedures.

Some of the topics that should be covered in an incident procedures plan include the following:

- *Introduction:* This section should contain a clear statement of project objectives, a description of the work to be done, desired timeframes, reasons the work is important, and other resource considerations.
- *Safety Plan:* The health and safety of all personnel MUST be of primary importance. Project managers MUST accept responsibility for preventing accidents and ensuring each employee works in a safe and healthy environment.
- Project managers MUST institute a policy to promote safe practices and to eliminate accidents and unsafe conditions. Trade-offs in safety procedures cannot be made in the interest of accomplishing work.
- A safety plan should include the following:
  - Project policies.
  - Goals and objectives.
  - Clear statement of responsibility.
  - Safety procedures.
- Accident prevention.
  - Job hazard analysis.
  - Orientation and training.

- Personal protective equipment.
- Hazard inspections.
- Air operations safety.
- Vehicle safety.
- Visitor safety.
- Traffic control road closures.
- First aid requirements.
- Safety meetings.

- Accident reporting and emergencies.
  - Forms and Guidelines.
  - Emergency medical plan.
  - Accident investigation.
  - Search and rescue.
- *Organisation:* Aerial seeding projects require experienced personnel with special skills. However, in many cases, these kinds of skills are not readily available and people are asked to perform tasks that may be unfamiliar to them. A clear description of the project organisation and the kinds of jobs to be done is needed.
- *Organisation chart:* Include a chart showing the incident organisation along with clearly defined lines of responsibility. Each person assigned to the incident should be shown as a box on the chart.
- *Job descriptions:* For each position on the chart, a written job description is needed. Job descriptions should include a general description of responsibilities, a detailed description of duties, and a listing of knowledge, skills, and abilities required by the position. Job descriptions are important because:
  - Give a clear picture of tasks to be done to both experienced and inexperienced personnel.
  - Provide basis for performance appraisals and evaluations.
  - Provide basis for developing future training programmes.
- *Communications:* The successful and safe completion of ESR incidents depends on good communications between all personnel involved. Adequate ground-to-ground and air-to-ground communications are essential. Someone that fully understands principles of radio communications and equipment involved needs to be assigned-to the project.
- The incident communications plan should include the following:
  - Dispatching procedures.
  - Downed aircraft reporting.
  - Radio frequencies.
  - Agencies frequencies.

  - Tactical frequencies.
- Operations.
  - Pre-project operations.
  - Secure landowner consent and other required clearances.
  - Designation and ground marking of seed block boundaries.
  - Designation of sensitive or no-treat areas.
  - Designation and approval of helispots and airstrips.
  - Hazard map.
  - Provide necessary training for incident personnel.
  - Incident orientation.
  - Aircraft safety.
  - Contract administration.
  - Calibration and characterisation of aerial application equipment.
  - Secure needed supplies and materials.
  - Locate and map locations of structural measures.
- During Operations.
  - Description of typical operations day.
  - Describe who is responsible for certain tasks at particular times during the day.
  - Personnel locations.
  - Communications checks.
  - GO/NO GO decision points.
  - Aircraft equipment checks.
  - Take off/landing records.
  - Load logs.
  - Accomplishment reporting.
  - Next day programme of work.
- Description of decision standards.
  - Weather
  - Wind
  - Moisture
  - Air turbulence
- Improper seed dispersal
  - Monitoring procedures
  - Correction of problems
  - Retreatment decisions
- Aircraft equipment problems
  - Mechanical problems
  - Seed delivery problems
  - Avionics

  - Personnel
- Required reports and records distribution.
  - Material transfer records
  - Daily accomplishment
  - Daily aircraft records
  - Contract daily diaries
  - Daily radio logs
  - Daily shift plans
  - Operations check list
  - Monitoring results
  - Tailgate safety meeting documentation
  - Post operations - Carry out demobilisation plan as described below.
- Demobilisation plan.
  - Debrief project personnel.
  - Performance appraisals.
  - Permanent project records file.
  - Accomplishments.
  - Cost accounting.
  - Operations plan.
  - Return equipment.
  - Project critique.
  - Letters of appreciation to project staff and other cooperators.
- Appendices.
  - Calibration formulas.
  - List of seed mixes and desired number of seeds per square foot.

PREPARING FOR AERIAL SEEDING PROJECTS

Burned Area Emergency Stabilisation and Rehabilitation (ESR) projects are EMERGENCY situations. Anything that can be done to facilitate the process prior to the need arising will most certainly be helpful. Because of the many complexities involved in implementing aerial seeding prescriptions, a project manager should not go into an incident cold. He or she will be much better off some preplanning has been done.

There are many tasks that can be completed ahead of an incident that will facilitate efficient implementation of aerial seeding prescriptions. These include but are not limited to the following: Agency Administrator Commitment - Brief your appropriate agency administrator on the objectives of ESR. Make sure they understand its EMERGENCY nature and the URGENCY of getting the job done. Obtain their pre-incident commitment to accomplishing the work. Give ESR work PRIORITY over other work items.

Aerial Application Contracts - Assemble sample aerial application contracts from previous incidents. When embarking on a project and time is short, it is best to have some idea of how to proceed. Talk to someone who has had experience; or ff you have been involved before, use those methods that have proven effective.

CONTRACTING AND PROCUREMENT

Work with your contracting and procurement sections to develop a list of potential aerial application contractors. Not all operators are the same and they may have differing amounts of support equipment available. It is beneficial to have this list in advance because of the short time available to solicit bids. Talk with someone who has had experience in aerial application in your area.

Develop a list of REPUTABLE seed suppliers. Seed suppliers are like any other businesses. Some are good and some are not so good. Try to find out who has provided the best level of service in your area. Find out if there have been problems with any of the suppliers in the recent past.

Seed Laws - Fully understand the laws governing the purchase and application of seed in your local area. This includes provisions for seed testing, labeling, and procedures for non-compliance. One of the worst things that can happen on a ESR project is to apply seed containing noxious weeds or other undesirable species. Every effort should be taken to ensure this does not happen. Warehouse Space - Know the availability of large rental warehouse space in your local area. If a large scale seeding effort is planned, a place to store seed out of the weather is needed. This is particularly true when seed is delivered in bags. Also, provisions need to be made for loading and unloading seed at all times of the day or night. Fork lifts, or other loading machinery, need to be planned for.

PERSONNEL

Develop a list of personnel experienced in aerial application. In order for implementation to proceed smoothly, access to skilled individuals on short notice must be available. These individuals are often assigned to incidents for more than 14 days.

The kinds of skills required on an aerial seeding project include the following:

- *Map Preparation:* The importance of high quality maps on an aerial seeding project cannot be over emphasised. Maps provide the necessary link between survey team and implementation team. They are also necessary components of seed procurement and aerial application contracts. Seed is not always cheap and it must be known in advance how many acres are to be treated with a particular seed mix. It is also critical that seed be applied to the proper treatment areas.

- *Air Operations:* All aerial seeding projects require the skills of an experienced air operations manager. This person or persons should be familiar with fixed-wing and/or helicopter operations. They should fully understand the capabilities of aircraft involved including payload, maneuverability, and navigation requirements. Proper air-to-ground and ground-to-ground communications are essential on aerial application projects. Air operations personnel need to familiar with communications equipment and procedures. Since this person will be overseeing aerial application operations, he or she must be familiar with FAA and internal agency aircraft safety rules and procedures. This person also needs to be familiar with other kinds of equipment used in aerial application of seed. This includes seed buckets or other types of seed delivery systems and their proper calibration, loading equipment such as preload chutes and bins, and seed storage and transportation.
- *Communications/Dispatch:* Proper communications are an essential component of a successful ESR incident. An experienced communications unit leader will be an important part of any ESR implementation organisation.
- *Administration:* His person should be familiar with agency accounting procedures for ESR, be able to keep accurate records, and understand procurement procedures and authorities.

PREPARING FOR AERIAL SEEDING PROJECTS (ON-INCIDENT)

Once assigned the responsibility of implementing an aerial seeding project, a number of factors must be considered when planning and organising the operation, and in developing contracts. An excellent means of getting an early start is to get involved with the ESR survey team. This will provide you with an understanding of the prescriptions being considered and important characteristics of the land area to be treated.

Some of the most important items that need to be considered include the following:

- Size of fire or complex of fires.
- Geographic location and accessibility.
- Terrain.
- Elevation and climate.
- Management areas/sensitive areas.
- Landowners involved.
- Number of seed mixes.
- Timeframes.

CALCULATING SEED RATES BASED ON PURE LIVE SEED PER SQUARE FOOT

Seeding rates for grasses are generally expressed in pounds per acres.

However, if a seeding prescription is based on pounds per acre instead of Pure Live Seed (PLS) per square foot, excess costs can be experienced. Generally, to achieve efficient erosion control, seeding rates should be in the range of 30 to 60 PLS per square foot.

Seeding rates can be calculated if you know the following:

- The total number of seeds per pounds
- The percentage of each pound that is PLS
- How many acres needing treatment
- The target PLS per square foot rate

Example:

Seed one acre with Standard Crested wheatgrass which has 175,000 seeds per pound and is 76 per cent PLS to get a result of 40 PLS per square foot.

(1 acre)*(43,560 ft2/acre)*(40)=1.742,000 PLS
(175,000)*(0.76)=133,000 PLS/lb
(1,742,400)/(133,000)=13.1 lb

Example:

Seed one acre with three species at different rates to obtain a coverage of 30 PLS/ft2

Species 1 = 175,000 seeds/lb; 76 per cent PLS; 30 per cent of mix
Species 2 = 645,000 seeds/lb; 71 per cent PLS; 40 per cent of mix
Species 3 = 227,000 seeds/lb; 87 per cent PLS; 30 per cent of mix
(1 acre)*(43,560 ft2/acre)*(30)=1,306,800 PLS

- Species 1: (1,306,800)*(0.30)=392,040 PLS
  (175,000)*(0.76)=1333,000 PLS/lb
  (392,040)/(133,000) = 2.95 lb
- Species 2: (1,306,800)*(0.40)=522,720 PLS
  (645,000)*(0.71)=457,950 PLS/lb
  (522,720)/;(457,950) = 1.14 lb
- Species 3: (1,306,800)*(0.30)=392,040 PLS
  (227,000)*(0.87)=197,490 PLS/lb
  (392,040)/(197,490) = 1.99 lb

Total Mix: 6.087 lb/acre

RESULT OF SEED DISPERSAL

Early summer rains in late June and early July effect a major contribution to the dispersal of saguaro seeds by dislodging seed masses still remaining on the plant, and by washing individual seeds into protected locations that provide concealment from predators and are favourable sites for subsequent germination and establishment. The early rains reduce the attrition of seeds and ordinarily set the stage for germination during subsequent summer rains. Such seed dispersal to more favourable sites accomplished by gravity and rainfall run-off is mainly effective for downslope dispersal in sloping habitats. Animal consumers, principally birds, which pass undigested viable seeds through their digestive tracts, are the primary agents for seed dispersal upslope and in flat habitats. Obligatory seed-eaters that efficiently digest consumed seeds, however, make little or no contribution to seed dispersal. Attrition. In the complete life cycle of the saguaro, the greatest mortality occurs during the pre-germination (seed) stage, the 1- to 5-week period between seedfall and the occurrence of germinating summer rains. During that period the major portion of the annual seed crop is consumed by animals—birds, mammals, and insects.

To some degree, the fruits of the saguaro are utilised either as food or a source of moisture by nearly every warm-blooded animal member of the community and by several species of insects. Initially heavy losses to feeding birds, especially doves, occur while fruits are still attached to the plant. The greatest losses, however, accrue after fruits drop to the ground where the diversity of consumer species ranges from ants to coyotes. The relative importance of consumer species varies according to their size and foraging efficiency, diversity of feeding habits, and relative abundance within the particular community.

Generally, obligatory seed-eaters—principally harvester ants, doves, and heteromyid rodents—exert the greatest impact as they are abundant and efficient consumers. However, in habitats where they are abundant (such as the Cactus Forest of the east monument), round-tailed ground squirrels consume the major portion of dropped fruits.

The impact of specific saguaro fruit consumers varies greatly from one desert community to another. In flat habitats of the east monument, relatively dense populations of harvester ants and round-tailed ground squirrels quickly remove most of the seeds which reach the ground. However, in flat habitats of the west monument where neither species is abundant, a large proportion of the seed crop remains on the ground, undisturbed throughout the pre-germination period. In some locations, seeds remain over winter on the ground. However, as a result of destruction by insects, microorganisms, or climatic action, few if any survive or contribute significantly to the following year's germinable seed supply. Most of the fruits drop to the ground immediately beneath the parent plant. Intensive animal activity sponsored by this abundance of food not only results in heavy attrition of seeds from these sites but also severely limits the suitability of such sites for germination and seedling survival, especially in flat habitats.

In view of abundant germination observed during favourable years, it is unlikely that the lack of young saguaros evident in some habitats is attributable to attrition of seeds.

Germination: The principal germination of saguaro seeds takes place in July and August from seeds of the current year's crop. Optimum conditions for natural germination coincide with the first full development of monsoon storms. Then, during the period from mid-July to mid-August, germination is normally associated with the occurrence of two or more rainstorms within a 2- to 5-day period.

The availability of moisture is the critical determinant in the germination process; temperature and light requirements are readily satisfied within all natural habitats of the saguaro. In continuous contact with free water, germination takes place in 48-72 hr.

The germination process is facilitated by the capability of seeds to absorb moisture hygroscopically. We conclude that this is an important adaptive strategy. Pre-conditioning by hygroscopic imbibition can reduce the critical period of required contact with free water by as much as one day, importantly mitigating the need for a prolonged period of saturation at the soil surface.

The saguaro is dependent in its germination requirements on a physically modified microenvironment produced by trees, shrubs, rocks, or other shade-producing objects. These shaded microhabitats provide moderated daytime temperatures that are within the upper range required for germination and prolong periods of high moisture availability at the soil surface. Ultimately, the number of adequately shaded sites suitable for saguaro seed germination is limited by the physical structure of the community.

Experimental evidence indicates that fewer than 1 in 200 seeds that reach a site where germination can occur survive to the seedling stage. We estimate that the net natural survival to the initial stage of seedling establishment is less than 1 per 1000 seeds produced.

The summer climatic environment of the Tucson area is well within the range necessary for saguaro seed germination; this exceeds, in germination suitability, the drier, more westerly portions of the species' range. At Tucson, depending upon the overall intensity of monsoon development, the year-to-year suitability of conditions for natural germination ranges from poor to near optimum, but some natural germination occurs in all years.

It seems likely that the western limits of saguaro distribution in California, Arizona, and northern Sonora are controlled by insufficient moisture for germination. Elsewhere to the northeast and south, it is clear that summer climatic environments would permit natural germination far beyond the present limits of saguaro distribution. Those limits, therefore, must be determined by factors that operate during the post germination stages of the plant's growth.

CHARLES DARWIN AND OCEAN DISPERSAL OF SEEDS

Plant dispersal by ocean currents has fascinated many famous explorers, including Charles Darwin and Thor Heyerdahl. Geodetic Survey using stoppered bottles containing a numbered postcard. When a bottle is found on a beach the finder fills out the card and drops it in the mail. It takes about one year for a drift bottle to float from Yucatan to Ireland. A bottle launched near Caracas, Venezuela reached the Florida Keys four months later, traveling at an average speed of 16 statute miles per day. It is estimated that tropical seeds found on European shores probably have been adrift for a year or longer.

During his famous voyage around the world on the H.M.S. Beagle, Charles Darwin championed the idea of drift seeds and fruits colonizing distant islands, particularly isolated volcanic islands that have never been connected to the mainland. Darwin studied the role ocean currents played in the flora of Cocos Keeling Islands in the Indian Ocean, and concluded that most of the endemic vascular flora was derived from drift seeds and fruits.

After he returned to England, Darwin conducted flotation experiments with cultivated plants. In the Journal of the Proceedings of the Linnaean Society Darwin stated: "I soon became aware that most seeds, in accordance with the common experience of gardeners, sink in water; at least I have found this to be the case, after a few days, with the 51 kinds of seeds which I have myself tried; so that such seeds could not possibly be transported by sea-currents beyond a very short distance."

Darwin also mentioned rafting as a dispersal mechanism for seeds that generally don't float well in sea water. In addition, he stated that seeds contained within pods, capsules and the heads of Asteraceae may be carried by ocean currents and washed ashore on distant beaches. In his Origin of Species, 1859, Darwin summarized his experimental data on seed dispersal in salt water, and expressed a higher confidence in dried seeds: "Therefore it would perhaps be safer to assume that the seeds of about 10/100 plants of a flora, after having been dried, could be floated across a space of sea 900 miles in width, and would then germinate."

Of all the 250,000 species of seed plants on earth, only about 250 species (0.1 per cent) are commonly collected as drift disseminules on tropical beaches; and only about half of these are known to produce seeds that can float in seawater for more than a month and still be viable. This relatively small number of drift seed species does not include seed plants which are dispersed on vegetation rafts, drift garbage from ships, or true marine seagrasses which live totally submersed in seawater. Although the total number of drift seed species with long viability periods may be relatively small, they nonetheless form a floral flotilla comprising countless thousands of individuals riding the ocean currents of the world.

The Hawaiian archipelago has been isolated from continental land masses during the past 30 million years, and yet the 1,000 species of indigenous Hawaiian angiosperms are believed to stem from natural introduction by long-distance dispersal of 280 ancestral plant about 14 per cent of the original flowering plant immigrants to the Hawaiian Islands are clearly adapted to oceanic drift. If dispersal by birds and air currents are ruled out, it appears that seeds were carried thousands of miles to these islands, possibly by rafting or within protective capsules and pods. For example, California tarweeds are not included with tropical drift seeds, and yet an ancestral tarweed traveled at least 3,000 miles to the Hawaiian Islands where it gave rise to a remarkable group of endemics known as the "Silver Sword Alliance." The small seeds from ancestral members of the lobelia family (Campanulaceae) also reached these islands giving rise to an unusual group of Hawaiian lobelioids. One of these is the pachycaul Brighamia insignis that grows on steep sea cliffs on the island of Kauai.

An interesting coral tree endemic to dry, leeward slopes of all the main Hawaiian islands is called wiliwili (E. sandwicensis). This Hawaiian endemic belongs to a group of closely related passerine species in Erythraster, including E. tahitensis, endemic to Tahiti; E. variegata, a widespread species in tropical Asia and the South Pacific; and E. velutina, another widespread species in northern South America, the West Indies and the Galapagos Islands. Although the seeds of wiliwili sink in water, it may have originated from an ancestral species with buoyant seeds that reached these islands many thousands of years ago. In fact, a related tropical Asian species (E. variegata), which is commonly planted as a street tree in Hawaii, has seeds that float in water. The seeds of this species sometimes show up along Hawaiian beaches and tidal waters; however, they are usually of local origin. Trees of E. variegata are fairly easy to identify because the pods are longer than most cultivated coral trees, often up to 12 inches or more in length.

. Although the reasons are not identical, the loss of flight in certain insects and birds. "If a plant shifts its ecological preference, it will tend to lose contact with the agent responsible for its dispersal."

Seeds of the Asian coral tree (Erythrina variegata) are buoyant in seawater and may have drifted to distant shores of the tropical Pacific. In fact, this species (or its progenitor) may have given rise to several species of endemic coral trees in the tropical Pacific region.

THE HAWAIIAN SILVER SWORD

The Hawaiian islands are thousands of miles from the nearest continent or island mass. Even though they include some of the most isolated islands on earth, they contain numerous endemic species of plants, many of which evolved from ancestral seeds that reached the islands by drifting. Some striking members of the sunflower family evolved on the Hawaiian Islands from an ancestral California tarweed (Hemizonia) that colonized these isolated Pacific islands millions of years ago.

This group of plants, called the Silver Sword Alliance, includes three genera and about 30 endemic species, an excellent example of adaptive radiation. One of the most amazing of all these Hawaiian endemics is the silver sword (Argyroxiphium sandwicense ssp. macrocephalum) that only grows in the cinders of Haleakala Crater on the island of Maui. The rosettes of sword-shaped leaves are covered with silvery hairs that reflect light and heat and provide insulation against the intense solar radiation and extreme aridity of this 10,000 foot (3,000 m) volcanic mountain.

A silver sword (Argyroxiphium sandwicense ssp. macrocephalum) in full bloom inside the lunar-like crater of Haleakala on the island of Maui. This remarkable plant and its relatives are descendents of an ancestral California tarweed (Hemizonia) that reached these islands millions of years ago, presumably by drifting.

NICKERNUTS

Some of the most ubiquitous drift seeds that readily germinate on tropical beaches are the gray and yellow nickernuts, Caesalpinia bonduc, C. ciliata, C. major. All are climbing, sprawling beach shrubs armed with vicious, recurved thorns and prickly pods bearing gray or yellow seeds. The smooth, marble-like seeds are commonly strung into necklaces and bracelets, often mixed with other colourful seeds. In Guayaquil, Ecuador, drilled nickernuts are sold by street vendors as amulets to ward off evil spirits. On some Caribbean islands nickernuts are used in a strategy board game called "Island Waurie."

Two players sit on opposite sides of a hardwood board with six depressions (wells) on each side, each well with four nuts. Through a series of carefully planned moves, players attempt to empty all six of their wells before their opponent can do so. Legend has it that Island Waurie was originally introduced to the Cayman Islands by Blackbeard the Pirate on one of his voyages from South Africa. The game became popular with Cayman Islanders and was enjoyed by Ernest Hemingway whenever he visited the islands. Know in Africa as "mancala," the games were played for thousands of years in Egypt, where boards have been found carved into the stone of the pyramid of Cheops and the temples at Luxor and Karnak.

THE COCONUT

Probably the best known of all plant drifters is the coconut (Cocos nucifera). In fact, it is hard to imagine a tropical beach without coconut palms. The origin of the coconut it has never been found truly wild, every coconut palm is planted by man or derived from such a planting." "There is no island or shore where its presence is not due directly or indirectly to its having been planted by man." Probably under most situations, coconut seedlings require watering and other attention from humans; cite several localities where coconuts appear to have seeded themselves naturally, including cays in British Honduras (Belize), rocky islets in the Fiji group, the east coast of Trinidad, Cocos-Keeling Atoll in the Indian Ocean, and Krakatau and adjacent islets following the catastrophic eruption of 1883.

Coconuts have naturally established themselves on beaches of the tropical Pacific. but they all belong to either of two major types known as niu kafa and niu vai. The niu kafa types have an elongate, angular fruit, up to 6 inches in diameter, with a small egg-shaped nut surrounded by an unusually thick husk. Niu vai types have a larger more spherical fruit, up to 10 inches in diameter, with a large, spherical nut inside a thin husk. The niu kafa type represents the ancestral, naturally-evolved, wild-type coconut, disseminated by floating. The niu vai type was derived by domestic selection for increased endosperm ("meat" and "milk") and is widely dispersed and cultivated by humans.

Both types of fruit can float, but the thicker, angular husk adapts the niu kafa type particularly well to remote atoll conditions where it can be found today. The presence of "undesirable" wild-type coconuts growing in mangrove swamps is clear evidence that they were self-sown and not planted by farmers. Angular, small-seeded coconuts on Ambergris Caye, off the coast of Belize. These coconuts resemble the wild-type niu kafa They were probably introduced to the Caribbean region of Central America by Portuguese traders presence of coconut palms on Cocos Island off the coast of Costa Rica and parts of the Pacific Coast at the time of Columbus. One of the proponents of the New World origin is Thor Heyerdahl, Heyerdahl expresses great confidence in Polynesian sailors who crossed vast stretches of the Pacific Ocean carrying staple foods from the New World, such as sweet potato, plantain and the coconut.

The sweet potato (Ipomoea batatas). Archaeological evidence shows that sweet potatoes were cultivated in South America by 2400 B.C. and fossilized sweet potatoes from the Andes have been dated at 8,000 to 10,000 years old. Although the sweet potato is clearly native to South America, it was cultivated in Polynesia as early as 1200 A.D. In fact, the sweet potato had already become the principle food in New Zealand by the time of Captain Cook's historic voyage to that part of the world in 1769. It is interesting to note that the sweet potato is known as "kumar" or "kumal" in the Lima region of coastal Peru, and it is called "kumaraHeyerdahl (1968) postulated that sweet potatoes were carried across the Pacific by Peruvian Indians before Europeans began to sail the world's oceans.

The coconut is indigenous to the Indo-Malaysian region. It spread by sea currents to many Pacific Island groups with adequate rainfall and moderate temperatures where the seedlings became established along beaches. Coconuts prefer the well-drained coral sand beaches of tropical islands and atolls, and are poorly established along shores of continents. Naturally dispersed coconuts can withstand occasional brief salt water flooding. Developing coconut palms obtain fresh water and mineral nutrients from a lens (soil layer) of fresh water (derived from island rainfall) which literally floats above the denser salt water beneath the beach sand.

3,000 miles seems to be the average maximum distance that a coconut will remain afloat and still remain viable. These limitations greatly diminish the chance of a coconut reaching the New World, let alone sprouting on a continental shoreproviding thatching, food, drink, coir, copra, oil, and hard endocarps which are fashioned into all sorts of decorative and useful articles.

The origin of the generic name for coconut "Cocos" may be traced to the three germination pores on the endocarp layer surrounding the seed. Portuguese and Spanish traders introduced the coconut into West Africa after 1500. They called it "coco" from the Portuguese or Spanish slang word for monkey face, supposedly because of the eye pattern on the endocarp and the brown, fibrous hair (husk). Coconuts were later introduced into the Americas by these early traders. The center of origin for ocean-dispersed coconuts appears to be the Indo-Malaysian region.

Perhaps the best physical evidence for an Indo-Pacific origin of the coconut comes from New Zealand and India, where fossil specimens of a species of Cocos have been fossil coconuts of an extinct Miocene palm (Cocos zeylandica) have been found in North Island, New Zealand. Fossil evidence from such far-flung areas the coconut had ample time to disperse naturally in the Indo-Pacific, before humans came along to speed up the process.

THE GALAPAGOS ISLANDS

The Galapagos Islands, always a provocative place for students of natural history, offer a wide variety of examples of drift dissemination, including the independent transport of seeds and fruits floating in the ocean and the transport of seeds and animals on vegetation rafts from the mainland. Several species of mangroves grow on these islands, including the ubiquitous red mangrove (Rhizophora mangle) with its conspicuous prop or stilt roots descending from its limbs. The cigar-shaped disseminule, known as a "sea pencil," is unusual because it is a germinated seedling and not a seed or fruit. The seedling, with its long, pendant taproot (mostly hypocotyl), drops from the parent plant, where it may be automatically planted in the mud of shallow water or washed away to a distant shore. Beach vines, such as beach morning glory (Ipomoea pes-caprae), gray nickernut (Caesalpinia bonduc) and beach bean (Canavalia maritima), occur on several of the islands. Their durable seeds ride the currents of the world's oceans, resulting in enormous pantropic distributions.

Drift seeds and fruits from the shores of Floreana Island in the Galapagos Archipelago. A. Nickernut (Caesalpinia bonduc), B. Manchineel (Hippomane mancinella), C. Beach Morning Glory (Ipomoea pes-caprae), D. Sea Purse (Dioclea sp.), E. Sea Bean (Mucuna sp.) and F. Prickly Palm (Acrocomia sp.). D, E, and F are mainland species that probably drifted to this island.

It would take about two weeks for a floating log or vegetation raft to reach the Galapagos Islands from mainland Ecuador. Considering that the archipelago is over 3 million years old, the drift seed scenario offers a plausible explanation for how some plants made the 600-mile (965-km) journey.

Successful dissemination by drifting would not have to occur often for it to have a profound effect on the ecology of the islands. Taking into account all methods of dispersal, including transport by wind, birds, and drifting on the ocean surface, Duncan Porter, a botanist at Virginia Polytechnic Institute and State University, estimates that 378 original introductions could account for the 522 indigenous plant species known to grow on the islands. Of these natural introductions, 59 per cent were a result of transport by birds, 32 per cent by wind transport, and 9 per cent by drifting on the ocean surface. If only one species successfully drifted to the islands and germinated there every 50,000 years, this would still account for the introduction of roughly 10 per cent of the islands's flora., it would require only one species to arrive and become established every eight to ten thousand years.

Ancestors of present-day Galapagos reptiles, such as the remarkable marine iguanas, may have also drifted to the islands. In the swift current of Ecuador's Guayas River, trees, branches, and large mats of vegetation broken loose from riverbeds are commonly carried out to sea. Reptiles now on the Galapagos may have ridden such rafts from the mainland many thousands of years ago. A marine iguana (Amblyrhynchus cristatus) on the rocky shoreline of Hood Island in the Galapagos Archipelago. These unusual sea-dwelling iguanas are descendants of an ancestral mainland species that rafted to these islands thousands of years ago. In the distance is the infamous M/N Bucanero.

THE PANTROPICAL MANGROVES

Mangroves include several dozen species of trees and shrubs that grow along tidal swamplands throughout tropical regions of the world. Since their roots are emersed in water-logged silt and mud, often deficient in oxygen, the buttressed trunks and aerial roots of mangroves are typically dotted with numerous lenticels—small pores which provide gas exchange between the roots and the atmosphere. Lenticels are especially conspicuous on the prop roots of the ubiquitous red mangrove (Rhizophora mangle) and on the deeply-fluted trunk of the tea mangrove (Pelliciera rhizophorae), an interesting swamp tree of the tea family (Theaceae) along the Pacific coast of Costa Rica. The lateral roots of some mangroves produce upright, slender outgrowths called pneumatophores which extend above the mud and water like snorkel tubes. The porous pneumatophores are covered with lenticels and provide additional aeration for the root systems. The black mangrove (Avicennia germinans), of Central America and the Caribbean region, produces literally hundreds of pencil-like pneumatophores around its base.

Mangroves survive in seawater with a salinity that would be lethal to most trees and shrubs. Like celery or carrot sticks placed in saltwater, the roots of most plants rapidly lose water if they are suddenly emersed in seawater. Halophytes (salt-loving plants), such as mangroves, generally have a lower concentration of water molecules (lower water potential) in their root cells so they can take in water. They maintain lower water potentials in their roots by having higher internal salt concentrations than seawater and by losing water at the leaf surface. Since high internal salt concentrations can be lethal to plant cells, some species such as the black mangrove and white mangrove (Laguncularia racemosa), can excrete excess salt through special glands in their leaf blades and petioles. Red mangroves have root cell membranes which prevent the absorption of excess salt.

The seeds of mangroves are especially remarkable because they commonly germinate within their fruit while still attached to the parent plant, a condition known as "viviparous seeds." Having their embryonic root (hypocotyl) already elongated gives them a better chance of establishing themselves in soft mud during low tide. Called "sea pencils," the cigar-shaped seedlings (disseminules) of red mangrove may have an elongate taproot up to 10 inches long when they drop from the parent tree. The unusual, top-shaped fruit of tea mangrove contains one of the largest seeds in the world (excluding palms).

It floats with the elongate, embryonic root pointing downward, and readily becomes implanted in soft mud. Germinated seedlings and sprouted fruits of the black mangrove and white mangrove are also dispersed by ocean currents. Red mangrove (Rhizophora mangle) with two viviparous (germinated) seedlings which have emerged from their fruits. The elongate embryonic axis called the hypocotyl develops into an extended taproot; Center: Onion-sized fruits of the tea mangrove (Pelliciera rhizophorae). The pointed beak encloses the embryonic root (hypocotyl). The fruit floats with the beak facing downward. When it becomes stranded in shallow water at low tide, the beak becomes implanted in the soft mud or silt, thus fascilitating the establishment of the seedling; Right: A sprouted fruit of black mangrove (Avicennia germinans) showing an elongated root.

Red mangrove (Rhizophora mangle) with a viviparous (germinated) seedling still attached to the fruit on the parent shrub. The seedling drops into the water where it is automatically planted in the soft mud or floats away.

POLYNESIAN BOX FRUIT

Beaches of French Polynesia are often littered with a buoyant drift fruit resembling a small coconut with flattened sides. It is called box fruit (Barringtonia asiatica) and is one of the most durable and widespread of all drifters, remaining buoyant for at least two years. In fact, they are used as fishing floats in Southeast Asia. Box fruits (Barringtonia asiatica) are widespread drift fruits in the tropical Pacific, remaining buoyant for more than two years. They are common in the turquoise-blue waters of French Polynesia

POLYNESIAN TAMANU

An equally common drift fruit of the tropical Pacific is called "tamanu"

(Calophyllum inophyllum) by the Polynesians. The smooth, gray fruits resemble ping-pong balls strewn along wave-swept beaches. This tree belongs to the garcinia family (Clusiaceae syn. Guttiferae). " common drifter, called tropical almond (Terminalia catappa), resembles an oversized unshelled almond. It is one of the most distinctive trees of tropical beaches with tiered, pagodalike limbs and colourful red and yellow leaves reminiscent of a deciduous forest in autumn. Seeds of the Polynesian "tianina" or lantern tree (Hernandia nymphaeifolia) are produced in fleshy red or white "lanterns" which float like colourful boats in the clear blue water. The woody, seed-bearing endocarps are polished by native islanders and made into shiny brown leis and necklaces.

The smooth, ping-pong ball fruits of tamanu (Calophyllum inophyllum) commonly drift ashore on beaches of French Polynesia. On the atoll of Tetiaroa they are strung on fishing line to make unusual lamp shades. Flowers of tamanu (Calophyllum inophyllum). The stamens occur in five bundles. On the small East-Indonesian island of Alor, this pantropic tree is called "camplung." Here the sticky seed within the dried pericarp is used for lighting. It burns with a clean flame and reportedly fends off mosquitoes.

THE POLYNESIAN PANDANUS

Another widespread Polynesian plant called screw pine (Pandanus tectorius) is also dispersed by ocean currents. The large multiple fruit (resembling a pineapple) is composed of buoyant, one-seeded sections called "keys." The hard, woody keys are also polished and made into necklaces and leis.

The multiple fruit of Pandanus tectorius, showing the individual one-seeded sections called "keys." In addition to the edible seeds (one inside each key), the keys are polished and used for necklaces and leis. The keys are very buoyant and water-resistant, and remain viable for months. They are dispersed by ocean currents to shores of distant atolls and islands throughout the tropical Pacific.

Sea Beans

Some of the most beautiful of all drift seeds are called "sea beans." They come from large climbing vines (called lianas) throughout rain forests of the New and Old World tropics. The seeds are produced in large pods and resemble round, flattened beans with hard, woody seed coats. In species of Mucuna the pods are covered with whiskerlike, stinging hairs—presumably to discourage ravenous seed predators. The pods develop from clusters of greenish, bat-pollinated blossoms at the end of long, ropelike branches which hang from the forest canopy. Sea beans are also called "hamburger seeds" because they have two rounded halves enclosing a conspicuous central connection layer (hilum). In Mexico and Central America they are also called "ojo de buey" because of their uncanny resemblance to the eye of a steer. Another species of sea bean (M. argyrophylla) called "ojo de venado" (deer eyes) has a most remarkable use in Central America. he sex of the seed is determined by whether they sink or float in water. Those that sink in water are called "hembras" (female) and those that float are "macho" (male). The hemorrhoidal treatment requires the sufferer to carry a "male" and a "female" seed in their back pocket. Some species of sea beans are called sea purses (Dioclea) because they are shaped like a purse, including a circular hilum along the edge that superficially resembles a zipper.

Velvety pods of the sea bean (Mucuna argyrophylla) hang from long, ropelike stems in the rain forests of Belize. The hard, black seeds are called "ojo de buey" (eye of the steer) and "ojo de venado" (eye of the deer) by local residents.

The Sea Heart

Perhaps the most remarkable of all drift seeds resemble large wooden hearts and are called "sea hearts" (Entada gigas). The heart-shaped seeds are produced in huge bean pods up to six feet long, the longest of any legume. In the rain-soaked tropical forest near Golfito, Costa Rica, enormous sea heart lianas twine through the forest canopy like a gigantic botanical boa constrictor. The vine is known locally as "escalera de mono" or monkey ladder because it provides a maze of arboreal thoroughfares for New World monkeys. Torrential rains wash the seeds into streams and rivers where they reach the sea, and perhaps eventually the shores of a distant continent. Sea hearts are highly prized by beach combers and make beautiful pendants when polished. Sea hearts and a similar rectangular Old World species (E. phaseoloides) were commonly used in Norway and northern Europe for snuffboxes and lockets. The seeds were cut in half, the contents removed, and the woody seed coats hinged together.

Sea hearts have a long and colourful history in fact and fiction. Early naturalists thought the unusual seeds came from strange underwater plants whose origin was shrouded in mystery. In England sea hearts were carried as good luck charms by sailors embarking on a long ocean voyage. It was thought that if sea hearts could survive a long and perilous journey across the ocean, perhaps they could also protect their owner. Christopher Columbus was fascinated with objects that drifted ashore on beaches of the Azores, off the coast of Portugal. It is said that a sea heart provided inspiration to Christopher Columbus and led him to set forth in search of new lands to the west. In fact, the sea heart is called "fava de Colom," or Columbus bean, by Portuguese residents of the Azores.

The Mary's Bean

Another drift seed called Mary's bean (Merremia discoidesperma) was a special find to pious beachcombers. Named after the Virgin Mary, it is also called crucifixion bean because of a cross etched on one side. The distinctive seeds are produced in papery capsules on a climbing tropical vine. The cross is actually an impression where the seed was attached inside the capsule. Because of a thick, woody seed coat and internal air cavities, the seeds are very buoyant and may drift for months or even years at sea. Historically, the unique seeds have been used for good luck charms and to ward off evil spirits. A woman in labor was assured an easy delivery if she clenched a Mary's bean in her hand. Seeds were handed down from mother to daughter as treasured keepsakes. To this day Mary's beans are occasionally sold by street vendors in Costa Rica.

Some drift seeds and fruits on the Caribbean shores of Costa Rica. A. Starnut Palm (Astrocaryum sp.), B. Mary's Bean (Merremia discoidesperma), C. Oxyrhynchus trinervias, D. Crabwood (Carapa guianensis), E. Prioria copaifera, F. Sea Bean (Mucuna sp.), G. Sea Coconut (Manicaria saccifera), and H. Calatola costaricensis.

In its native habitat, the Mary's bean is known from only a few locations in Mexico and Central America. It is a lovely vine of the Morning Glory Family (Convolvulaceae) with beautiful yellow, funnel-shaped blossoms. As drift seeds, the Mary's bean is known from the Marshall Islands to beaches of Norway, a total distance of more than 15,000 miles (24,000 km). this constitutes the widest drift range of any seed or fruit which has been documented. Other pantropical drift seeds may drift as far or farther, but their precise point of origin cannot be determined with certainty.

Large Neotropical Drift Fruits

With the exception of introduced cattle, donkeys and horses, no native mammals of the New World tropics can crush the hard, thick-walled pods of many rain forest trees in their jaws. Livestock apparently like the sweet pulp inside the pods of West Indian locust (Hymenaea courbaril) and disperse the hard, viable seeds in their excrement. In areas without livestock, the rotting pods litter the ground beneath large trees. Agoutis, tapirs and peccaries chew open the rotting pods and eat the sweet pulp and seeds, but are not major agents of seed dispersal like the larger hoofed mammals. large grazing mammals, including extinct pleistocene elephants called gomphotheres, may have once eaten the pods and dispersed the seeds in lowland forests. In Africa, the large woody pods of related species are quickly devoured by large herbivores.

There are other Central American rain forest trees that also appear to be missing their natural herbivorous dispersal agents. Their hard, woody, indehiscent fruits pile up beneath the branches and slowly rot away in the soggy, moldy layer of soil and debris. The lack of natural dispersal agents may also apply to devil's claw pods that litter the ground in temperate regions of the New World. Hard, woody fruits of some Central and South American rain forest trees were likely dispersed by large prehistoric herbivores thousands of years ago. A. Cassia grandis (legume family—Fabaceae); B. Crescentia alata (bignonia family—Bignoniaceae); C. Hymenaea courbaril (legume family—Fabaceae); and the palms (Arecaceae): D. Attalea speciosa (one fruit cut open to reveal thick, woody pericarp); E. Raphia taedigera; and F. Orbignya cohune.

The Manchineel Tree

There are dozens of unusual drift seeds that wash ashore on Caribbean islands, each with fascinating stories about them. One of the most interesting is an intricately-sculptured star that comes from the fruit of a native tree called the manchineel tree (Hippomane mancinella). The fruit of the manchineel tree resembles a small green apple, except this "apple" has a woody, seed bearing endocarp and is quite poisonous. Manchineel trees are common along Caribbean shores, and the fruits often fall into the water. The outer fleshy layer rots away and the buoyant, woody endocarp starts a new career as a drift disseminule. After many months of wave action and abrasion in tropical waters, the endocarp gradually becomes eroded into an intricate star. It is a beachcomber's delight to find one of these beautifully-sculptured stars. Fruits of the manchineel tree (Hippomane mancinella) resemble small green apples. The seed-bearing pits (endocarps) float in tropical waters and gradually become eroded into intricately-sculptured stars.

The manchineel tree belongs to the diverse Euphorbia Family (Euphorbiaceae), and like many members of this family, it contains a poisonous milky-latex sap. The toxin is a mixture of diterpene esters, and contact with the skin may cause inflammation and a blistering rash. The caustic terpenes are structurally similar to phorbol, a potent carcinogen listed among organic There were wild fruits of various kinds, some of which our men, not very prudently, tasted; and upon only touching them with their tongues, their countenances became inflamed, and such great heat and pain followed, that they seemed to be mad, and were obliged to resort to refrigerants to cure themselves."

A young manchineel tree (Hippomane mancinella) that has sprouting in the sand of a Costa Rican beach. The beach was littered with the green, apple-like fruits of this ocean-dispersed species.

Index